Energy, Society and Environment

Technology for a sustainable future

Routledge Introductions to Environment Series

Published and Forthcoming Titles

Titles under Series Editors:
Rita Gardner and Antoinette Mannion

Environmental Science texts

Environmental Biology
Environmental Chemistry and Physics
Environmental Geology
Environmental Engineering
Environmental Archaeology
Atmospheric Systems
Hydrological Systems
Oceanic Systems
Coastal Systems
Fluvial Systems
Soil Systems
Glacial Systems
Ecosystems
Landscape Systems

Titles under Series Editors:
David Pepper and Phil O'Keefe

Environment and Society texts

Environment and Economics
Environment and Politics
Environment and Law
Environment and Philosophy
Environment and Planning
Environment and Social Theory
Environment and Political Theory
Business and Environment

Key Environmental Topics texts

Biodiversity and Conservation
Environmental Hazards
Natural Environmental Change
Environmental Monitoring
Climatic Change
Land Use and Abuse
Water Resources
Pollution
Waste and the Environment
Energy Resources
Agriculture
Wetland Environments
Energy, Society and Environment
Environmental Sustainability
Gender and Environment
Environment and Society
Tourism and Environment
Environmental Management
Environmental Values
Representations of the Environment
Environment and Health
Environmental Movements
History of Environmental Ideas
Environment and Technology
Environment and the City
Case Studies for Environmental Studies

Routledge Introductions to Environment

Energy, Society and Environment

Technology for a sustainable future

David Elliott

London and New York

First published 1997
by Routledge
11 New Fetter Lane, London EC4P 4EE

Simultaneously published in the USA and Canada
by Routledge
29 West 35th Street, New York, NY 10001

Typeset in Times by Keystroke, Jacaranda Lodge, Wolverhampton
Printed and bound in Great Britain by Biddles Ltd, Guildford and King's Lynn

British Library Cataloguing in Publication Data
A catalogue record for this book is available from the British Library

Library of Congress Cataloguing-in-Publication Data
Elliott, David
Energy, society, and environment : technology for a sustainable
future / David Elliott.
p. cm. — (Routledge introductions to environment)
Includes bibliographical references and index.
1. Power resources—Environmental aspects. 2. Technology—Social
aspects. I. Title. II. Series
TD195.E49E42 1997
333.79′14—dc20 96–41364

ISBN 0–415–14506–6 (hbk)
0–415–14507–4

For Oliver

Contents

Series editors' preface *Environment and Society titles*

The 1970s and early 1980s constituted a period of intense academic and popular interest in processes of environmental degradation: global, regional and local. However, it soon became increasingly clear that reversing such degradation would not be a purely technical and managerial matter. All the technical knowledge in the world does not necessarily lead societies to change environmentally damaging behaviour. Hence a critical understanding of socio-economic, political and cultural processes and structures has become, it is acknowledged, of central importance in approaching environmental problems. Over the past two decades in particular there has been a mushrooming of research and scholarship on the relationships between social sciences and humanities on the one hand and processes of environmental change on the other. This has lately been reflected in a proliferation of associated courses at undergraduate level.

At the same time, changes in higher education in Europe, which match earlier changes in America, Australasia and elsewhere, mean that an increasing number of such courses are being taught and studied within a framework offering maximum flexibility in the typical under-graduate programme: 'modular' courses or their equivalent.

The volumes in this series will mirror these changes. They will provide short, topic-centred texts on environmentally relevant areas, mainly within social sciences and humanities. They will reflect the fact that students will approach their subject matter from a great variety of different disciplinary backgrounds; not just within social sciences and humanities, but from physical and natural sciences too. And those students may not be familiar with the background to the topic, they may or may not be going on to develop their interest in it, and they cannot automatically be thought of as being at 'first-year level', or second- or third-year: they might need to study the topic at any year of their course.

The authors and editors of this series are mainly established teachers in higher education. Finding that more traditional integrated environmental studies or specialised academic texts do not meet their requirements, they have increasingly met the new challenges caused by structural changes in education by writing their own course materials for their own students. These volumes represent, in modified form which all students can now share, the fruits of their labours.

To achieve the right mix of flexibility, depth and breadth, the volumes, like most modular course themselves, are designed carefully to create maximum accessibility to readers from a variety of backgrounds. Each leads into its topic by giving adequate introduction, and each 'leads out' by pointing towards complexities and areas for further development and study. Indeed, much of the integrity and distinctiveness of the Environment and Society titles in the series will come through adopting a characteristic, though not inflexible, structure to the volumes. Each introduces the student to the real-world context of the topic, and the basic concepts and controversies in social science/humanities which are most relevant. The core of each volume explores the main issues. Data, case studies, overview diagrams, summary charts and self-check questions and exercises are some of the pedagogic devices that will be found. The last part of each volume will normally show how the themes and issues presented may become more complicated, presenting cognate issues and concepts needing to be explored to gain deeper understanding. Annotated reading lists are important here.

We hope that these concise volumes will provide sufficient depth to maintain the interest of students with relevant backgrounds, and also sketch basic concepts and map out the ground in a stimulating way for students to whom the whole area is new.

The Environment and Society titles in the series complement the Environment Science titles which deal with natural science-based topics. Together this comprehensive range of volumes which make up the Routledge Introductions to Environment Series, will provide modular and other students with an unparalleled range of perspectives on environmental issues, cross referencing where appropriate.

The main target readership is introductory level undergraduate students predominantly taking programmes of environmental modules. But we hope that the whole audience will be much wider, perhaps including second- and third-year undergraduates from many disciplines within the social sciences, science/technology and humanities, who might be taking occasional environmental courses. We also hope that sixth-form teachers and the wider public will use these volumes when they feel the need to obtain quick introductory coverage of the subject we present.

David Pepper and Phil O'Keefe
1997

Series International Advisory Board

Note on the text

Bold is used in the text to denote words defined in the Glossary. It is also used to denote key terms.

Figures

Tables

Boxes

Author's preface

Energy use is fundamental to human existence and it should come as no surprise that the way mankind has been using it is at the heart of many of the environmental problems that have emerged in recent years. There are many types of pollution, but the emissions from the combustion of fuels in power stations and cars are probably the most worrying for many people, given the impact of air quality on health. More generally, the use of fossil fuels such as coal, oil and gas is increasingly seen as having major environmental impacts, such as global warming. There are also major concerns over the use of nuclear fuels. These energy sources also underpin what many see as an unsustainable form of industrial society, in which the environment is treated as a 'free' resource of energy and other materials and as a more or less infinite sink for wastes, with the scale and pace of the 'throughput' from source to sink growing ever greater.

This book explores the way in which energy use interacts with society and the environment, but the emphasis is not so much on describing the problems as on looking at some possible strategic solutions. Some of the solutions involve new technology: our use of technology has been a major cause of environmental problems, but it is sometimes argued that technology can be improved and used more wisely so as to avoid them in future. However, there are also limits to what technology can do: in the end it may be that there will also be a need for social, economic and political change if serious environmental damage is to be avoided. Part of the aim of this book is to try to explore just how far purely technical solutions can take us, and then to look at the strategic alternatives.

This is not a source book of environmental problems or technical solutions. The emphasis is mainly on social processes and strategic issues, rather than on technical details. I have however provided some technological background where it is needed. There are also Further Reading guides at the end of each chapter, pointing to key texts, including those in the rest of this series, some of which look at the technological aspects in more detail. Equally, this is not a social science textbook. I have located the discussion of social issues within a technological context, through case studies and examples, to try to make the issues more concrete, and to illustrate the interaction between the technological and social aspects. I have also included some questions in the Appendices to help you consolidate and develop your understanding of the arguments.

The structure of the book

This book is structured in four main parts. In the first part there is a general introduction to the key *environmental* issues and problems. Then, in Part 2, there is a review of some of the key *technological* solutions, followed, in Part 3, by a review of some of their implementation *problems*. Finally, in Part 4, there is a discussion of the wider implications for *society* of attempting to develop a sustainable approach to energy use.

The first task, in Part 1, is to get an idea of the problems the world faces, by looking at the way energy is used and at the way energy use and the environment interact. Energy use is of course primarily related to other human activities – such as heating homes, moving objects and people, manufacturing things and growing food. In each case technologies have been developed which use fuels to provide power. So the initial survey in Chapter 1 looks quite broadly at the technology used and at the forces shaping its development, and sets out a conceptual model of interactions between technology, society and the environment.

Then Chapter 2 looks at the environmental implications of the use of technology, with energy use as the key issue. This analysis of the environmental problems associated with existing forms of energy generation and use leads on, in Chapter 3, to a basic set of *criteria* for environmentally sustainable energy technology.

With the survey of problems as a background, and armed with the criteria for sustainable technology, Part 2, looks critically at some possible technological solutions to energy-related environmental problems, some of which are seen as relatively limited 'technical fixes', in that they only deal with symptoms rather than causes. For example, many technical fixes are essentially post hoc 'remedial' measures, attempting to 'clean up' harmful emissions – the so-called 'end-of-pipe' approach. Thus flue-gas scrubbing devices are used to filter out harmful emissions from power stations and catalytic converters are added to the end of car exhaust pipes.

There are obviously limits to this sort of approach: some emissions cannot be easily or economically filtered out, notably the carbon dioxide gas which is a fundamental product of combustion and a major contributor to the greenhouse 'global warming' effect. Some fossil fuels produce less carbon dioxide than others, so there is some potential for reducing emissions by switching fuels, but in the end the only real way to reduce carbon dioxide production is to burn less fossil fuel. One response is therefore to try to reduce demand for energy by avoiding waste through the adoption of energy conservation measures at the point of use, such as insulation in buildings, and by developing more efficient energy using technology. In the last few years there have also been moves to develop 'greener' products and 'cleaner' production processes, and one of the aims has been to improve the efficiency with which they use energy. Some people believe that a shift to a 'conserver society' is a vital element in any attempt to live in an environmentally sustainable way and Chapter 4 looks at the basic technical elements of this approach.

However, although energy conservation has enormous potential, there will still be a need to generate some energy. Assuming that the use of fossil fuel must be reduced and perhaps eventually eliminated, the main alternative new energy supply options are nuclear power and renewable energy technology – energy derived from natural sources like the wind, waves and tides. Chapter 5 looks at nuclear power. Once seen as the energy source of the future, the technical, economic, environmental and social problems now seem to have multiplied, while the attractions of the alternatives to nuclear power – such as the use of natural renewable energy sources like wind – seem to have increased. The conclusion would seem to be that nuclear power is unlikely to meet our criteria, while renewable energy, if coupled with energy conservation, seems likely to be a viable option – at least technically. Chapter 6 therefore looks at the renewables in detail, reviewing the basic technology. Chapter 7 then looks at renewable energy developments around the world and Chapter 8 rounds off our survey of renewables with a discussion of development strategy.

Having set out the potentially sustainable energy supply option of renewable energy, Part 3 attempts to look at the problems relating to the deployment of renewable energy and other sustainable energy technologies. Chapters 9 and 10 look at some of the institutional obstacles and implementation problems, while Chapter 11 presents a case study of public reactions to the deployment of wind farms in the UK. This case study is analysed in Chapter 12.

Part 4 widens the focus and asks, assuming that the obstacles discussed in Part 3 can be overcome, to what extent can the various solutions discussed in Part 2 contribute to genuine environmental sustainability. Chapter 13 asks can we just rely on technical fixes or are social and political changes also needed? Chapter 14 reminds us that a global perspective is needed, and that this raises issues concerning world economic development.

Clearly our discussion must touch on some very broad and potentially contentious issues concerning, for example, consumerism, economic redistribution and political power, which no one book can hope to resolve. However, Chapter 15 attempts to round off our analysis by looking at the tactical and strategic choices for society, technology and the environment, while Chapter 16 summarises the overall conclusions and looks at some possible ways ahead.

Study notes

It is assumed that some readers will be using this book as part of a study programme in the context of higher education. These study notes are designed to provide an introductory educational guide.

Aims

The Preface sets out the overall rationale and structure of the book, but students may find it helpful to have a more formal checklist of aims and objectives.

Basically, **Part 1** of this book should help you to develop an overview of energy-related environmental issues, including an appreciation of how they have emerged historically. More specifically Part 1 should help you:

- to appreciate that the interaction between the various parts of human society and the rest of the ecosystem is complex and that human energy use plays a major part in shaping this interaction (Chapters 1 and 2).
- to understand the arguments for the adoption of more sustainable approaches to energy generation and use, and for the development of criteria for selecting the appropriate technologies (Chapter 3).

Part 2 is designed to help you to appreciate the strengths and weaknesses of the various sustainable energy options, including energy conservation and the 'alternative' (i.e. non-fossil) energy supply options. More specifically it should help you:

- to appreciate the potential and limitations of 'technical fixes' as a means of resolving environmental problems, and the need for more radical solutions (Chapter 4).
- to analyse the pros and cons of nuclear power (Chapter 5).
- to understand the nature and potential of renewable energy technology (Chapter 6).
- to appreciate some of the ways in which renewable energy has been developed and deployed around the world (Chapter 7).
- to understand some of the strategic development issues facing sustainable energy technology (Chapter 8).

Part 3 is designed to help you explore some of the institutional and social problems facing the development and use of sustainable energy technology. More specifically it should help you:

- to appreciate the institutional problems facing novel energy technologies seeking initial research support, e.g. from governments (Chapter 9).
- to appreciate the institutional problems facing novel energy technologies in obtaining finance for large-scale deployment (Chapter 10).
- to understand the problem of winning public acceptance for novel technologies, as exemplified by local reactions to wind farms (Chapter 11).
- to appreciate the need to negotiate trade-offs between the 'local costs' and the 'global benefits' of renewable energy technologies, in the context of winning public acceptance for them (Chapter 12).

Part 4 is designed to broaden the discussion to help you understand some of the key strategic issues facing sustainable energy technology. More specifically it should help you:

- to appreciate that there are differing views about what sustainability is, and whether it is needed or obtainable (Chapter 13).
- to understand some of the ways in which technological changes, the industrialisation process and economic development patterns have interacted historically around the world; and some of the environmental limits which might constrain continued economic growth (Chapter 14).
- to appreciate that there are no unique blueprints for sustainability, but that the future has to be negotiated to try to avoid inequitable developments and devise an acceptable pattern of use of technology. (Chapter 15)
- to appreciate how attempts have been made to develop and negotiate criteria for sustainability, and the various strategies that have emerged for bringing about change.

Learning aids

I assume that you will read through this text sequentially. But to aid your navigation around the text and to help you consolidate your understanding, as well as the index, I have provided 'topic summaries' at the beginning of each chapter and 'summary points' at the end. There are also some general questions (and answers) and a glossary of key terms at the end of the book. A list of abbreviations is provided on page xxvi.

Non-technical readers will hopefully find the introduction to basic energy terms and units in Chapter 2 helpful, while the model of social–environmental interactions in Chapter 1 may be useful for non-social science students. I refer back to it regularly in the rest of the book, particularly in Part 4.

Preliminary questions

As a preliminary exercise before starting the book, you might like to consider the following questions. Some short notes on them are also included, but write down your reactions to the questions before looking at the notes. You might like to come back to the questions, the notes and your answers after you have finished the book, to see to what extent you have developed your understanding, or changed your views.

Questions

1 To what extent does your own personal lifestyle affect the planet in terms of the energy you use at work and play?
2 To what extent could you reduce any adverse environmental impacts from your use of energy?
3 What do you think will be the key energy sources of the future?
4 Can the whole world adopt Western-style consumption levels (i.e. of goods and services)?
5 Are you optimistic about the future or do you fear that environmental problems will get worse?
6 What actions do you think can help secure a sustainable future?

Notes

1 As individuals, at least in the developed countries, we benefit from the use of electricity, gas and petrol but we are becoming aware that this is

causing environmental problems. Domestic heating and car use, along with air transport, are the major 'personal' uses of energy, but the provision of goods and services also involves very significant energy use, over which individual consumers have little influence.

2. Individual lifestyle changes can help reduce energy use (e.g. less car use), as can investing in more energy efficient homes and domestic devices. But to have a major impact on energy use there will have to be wider changes – in energy generation technology, industry and possibly also in society.
3. The fossil fuels (coal, oil, gas) will not last for ever. Nuclear power and natural renewable energy sources are the main alternatives – along with energy conservation. However, not everyone is happy to consider nuclear power as a safe option, and there may be problems with relying on renewables.
4. Industrialisation of the sort pioneered in the West is unlikely to be environmentally sustainable on a global scale. The developing and developed countries may have to adopt alternative technological approaches – and even then it is not clear if economic growth of the current sort can continue worldwide.
5. It is not clear whether humanity can adapt quickly enough in order to develop sustainable approaches to living on this planet – and some say we do not need to. But should it be thought necessary, the opportunity for change exists, assuming vested interests in the status quo can be overcome.
6. There is no one single way forward. Diversity is a good ecological principle, but if we wish to avoid destructive competition for scarce resources we will have to learn how to co-operate and negotiate more effectively.

Further study

This is only an introductory book, and many of the ideas and issues have inevitably been simplified. There are guides to further reading at the end of each chapter and a more general guide to further reading (Appendix III) at the end of the book. There are also extensive references within the text. I have provided cross-references to other books in this series: the two most relevant texts are Paul Hooper's *Environment and Technology* and Gavin Gillmore's *Energy Resources*. As you will find, I have tried to avoid overlap with them. I have also provided a list of contacts (Appendix II) should you wish to follow up specific initiatives or issues.

Facts and values

Finally, a word of warning. While I have tried to back up the arguments and analysis in this book with references, and to present the factual material as objectively as possible, some of the issues discussed are controversial. I have tried to strike a balance between extreme views and complacent views, and between optimistic and pessimistic interpretations, but inevitably biases will creep in. One of the skills you will need to develop in approaching this subject is to read critically and, so far as is possible, distinguish between facts and value judgements.

While I hope that I have presented a reasonably balanced account, not everyone will agree. This book does after all discuss challenges to the status quo. Some people may argue that the analysis is too radical or too partisan. Others may suggest that it is not radical enough. You will have to make up your own mind – and one way to try to do that is by reading other books written from other perspectives. That is why I have tried to ensure that some divergent views are represented in the Further Readings.

Acknowledgements

An early version of parts of Chapter 3 appeared in a paper I co-authored with Alexi Clarke in the *International Journal of Global Energy Issues* (9 (4/5), 1996). Chapter 11 is based on parts of a paper I produced for the House of Commons Welsh Affairs Select Committee (Second Report, Session 1993–4, vol. III, pp. 432–42, HMSO, London, 1994), a revised version of which was subsequently published as 'Public reactions to windfarms: the dynamics of opinion formation', in *Energy and Environment* (5 (4): 343–62, 1994). Chapters 7, 9, 10 and14 draw on material I developed for Blocks 5 and 7 of the OU course on Innovation: Design, Environment and Strategy (T302), 1996.

Versions of some of the material in this book have also appeared in *Renew*, the journal I edit for NATTA.

Thanks are due to Dave Toke, who commented on an early draft of this book, and to Tam Dougan for her critical reviews. She and our son Oliver not only coped with my preoccupation while writing this book but also provided much of the motivation.

Sally Boyle from the OU Design Discipline, aided by Richard Hearne, then provided the necessary skills to convert the graphics into understandable form. Sarah Lloyd and the team at Routledge converted the text into what you have now.

However, responsibility for the final product rests with the author.
In writing this book, designed to be accessible to non-technical people, I have inevitably had to try to simplify some complex technological issues. I trust practitioners will bear with me on this, although I would

welcome comments and criticism. Although I have tried to be evenhanded and to subject all the technologies and policies discussed to critical assessment, inevitably biases exist. This seems unavoidable if one is attempting to discuss alternatives to the status quo, but I would welcome comments on any part of the analysis that is felt to be misleading or incorrect.

Abbreviations

CCGT	combined cycle gas turbines
CEC	Commission of the European Community
CHP	combined heat and power (co-generation)
CPRE	Council for the Protection of Rural England
DNC	declared net capacity
DTI	Department of Trade and Industry
EC	European Commission (of the European Union)
EST	Energy Saving Trust
ETSU	Energy Technology Support Unit
FBR	fast breeder reactor
GW	gigawatt (1,000 megawatts)
ICT	information and communication technologies
IPCC	Intergovernmental Panel on Climate Change
kWh	kilowatt hour (1,000 watts for one hour)
MW	megawatt (1,000 kilowatts)
NFFO	Non-Fossil Fuel Obligation
PV	photovoltaics (solar cells)
PWR	pressurised water reactor
REC	regional electricity company
SRC	short rotation coppicing
TWh	terawatt-hour (1,000,000,000 kWhs)
UKAEA	United Kingdom Atomic Energy Authority

Part 1 Environmental problems

Part 1 reviews the way in which mankind's use of energy impacts on the environment. It looks at how energy use has grown dramatically since the industrial revolution and at some of the key environmental problems that have emerged. It also introduces the idea of sustainable development as an alternative approach and sets out some basic environmental criteria for sustainable energy technologies.

1 Technology and society

- **Energy use in society**
- **Environmental impacts**
- **Sustainable development**
- **Negotiating interactions**

This introductory chapter sets the scene by looking at the interaction between people and the planet, with the focus on energy use. The ever-increasing pattern of energy use seems unlikely to be environmentally sustainable, in which case we will need to try to negotiate a new way forward. To try to describe some key features of the human–environmental interaction, and how it might be modified, the chapter introduces an analytical model of the various conflicting interests, which is used throughout the book.

People and the planet

Human beings have developed a capacity to create and use tools – or what is now called technology. Technology provides the means for modifying the natural environment for human purposes – providing basic requirements such as shelter, food and warmth, as well as communications, transport and a range of consumer products and services. All of these activities have some impact on the environment. The sheer scale of human technological activity puts an increasing stress on the natural environment to the extent that it cannot absorb our wastes, while our profligate lifestyles lead us increasingly to exploit the planet's limited resources.

Energy resources are an obvious example of limited resources whose use can have major impacts. Figure 1.1 shows the gigantic leap in energy use since the industrial revolution. Certainly energy use is now central to

Figure 1.1 ***Growth in total fossil fuel consumption worldwide since the industrial revolution (in billion tonnes of oil equivalent)***

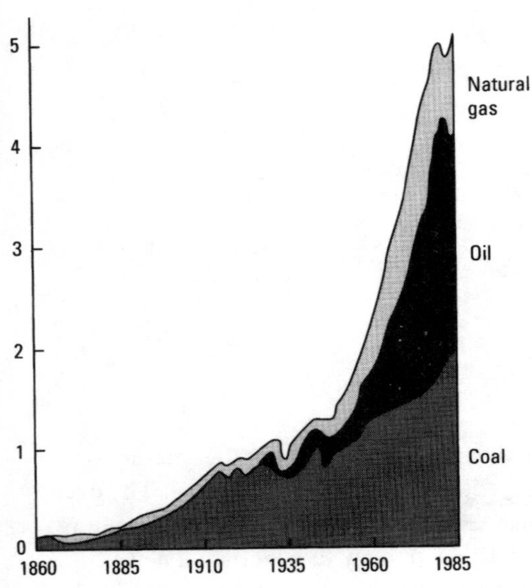

Source: *Physics Review* 2 (5), May 1993, based on data from D. A. Lashof and D. A. Tirpak (eds), *Policy Options for Stabilizing Global Climate*, US Environmental Protection Agency, Draft Report to Congress, Washington DC, 1989.

most human activities and many of our environmental problems could be described in terms of our energy getting and energy using technologies. The most obvious environmental effects are the physical impacts of mining for coal and drilling for oil and gas, and distributing the resultant fuels to the point of use. However, increasingly it is the use of these fuels that presents the major problems. Burning these fuels in power stations to generate electricity, or in homes to provide heat, or in car engines to provide transport, generates a range of harmful gases and other wastes and also, inevitably, generates carbon dioxide, a gas which plays a key role in the greenhouse 'global warming' effect. Figure 1.2 shows the seemingly inexorable rise in global carbon dioxide emissions, while Figure 1.3 shows the rise of planetary average temperature.

In the 1990s the upward trend in global average temperatures has continued. For updates see the World Wide Web pages posted by the Climatic Research Unit at the University of East Anglia: http://www.cru.uea.ac.uk/cru/press/pj9601/index.htm.

If this trend continues, the world climate could be significantly changed, leading, for example, to the melting of the ice caps, serious floods, droughts and storms, all of which could have major impacts on the ecosystem and on human life. Of course there remain many uncertainties over the nature and likely future rate of climate change. Nevertheless, in simple terms it seems obvious that something must be altered if significant amounts of the carbon dioxide are released into the atmosphere. This gas was absorbed from the primeval carbon dioxide rich atmosphere and was trapped in underground strata in the form of fossilised plant and animal life. We are now releasing it by extracting and burning fossil fuels. It took millennia to lay down these deposits, but a large proportion of these reserves may well be used up, releasing trapped carbon back into the atmosphere, within a few centuries.

Figure 1.2 ***Atmospheric concentration of carbon dioxide 1750–1988 (in parts per million)***

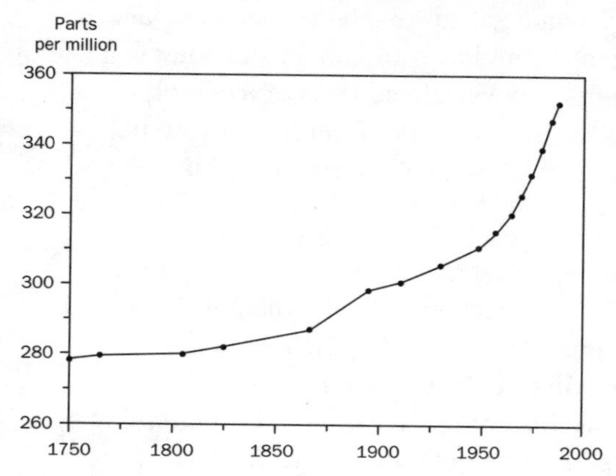

Source: C. Flavin, 'Slowing global warming: a worldwide strategy',Worldwatch Paper 91, Worldwatch Institute, Washington DC, 1989, based on data from Neftel and Keeling.

Figure 1.3 ***Global average temperature 1880–88 (in degrees Celsius)***

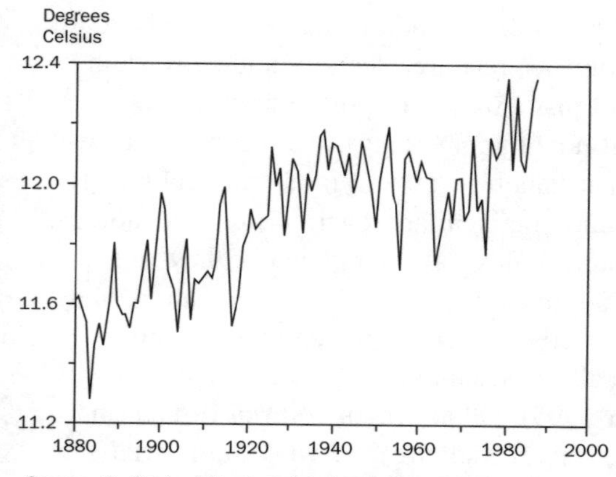

Source: C. Flavin, 'Slowing global warming: a worldwide strategy',Worldwatch Paper 91, Worldwatch Institute, Washington DC, 1989, based on data from Hansen.

In addition to major global impacts such as this there are a host of other environmental problems associated with energy extraction, production and use – acid emissions from the sulphur content of fossil fuels being just one. Air quality has become an urgent issue in many countries, given the links it has to health. The release of radioactive materials from the various stages of the nuclear fuel cycle represents an equally worrying problem, even assuming that there are no accidents. Accidents can happen in all industries – and they present a further range of environmental problems, the most familiar being in relation to oil spills.

Concerns over environmental pollution issues such as these first emerged on a significant scale in the 1960s and 1970s, in part following some spectacular oil spills from tankers, including the *Torrey Canyon* off Cornwall in 1967 and the *Amoco Cadiz* off Brittany in 1979. However the main environmental concern was over longer term strategic problems: in the mid-1970s, following a series of long-range energy resource predictions, it was felt that some key resources might run out in the near future (Meadows *et al.* 1972).

Certainly, a substantial part of the world's fossil fuel reserves have been burnt off and a substantial part of the world's uranium reserves have been used, although, as discussed in Chapter 3, there are disagreements about precisely what the reserves are, and in the 1980s and 1990s resource scarcity is seen as a less urgent problem. The more important strategic question nowadays is whether what resources are left can be used safely.

Sustainability

The basic issue is one of environmental sustainability: can the planet's ecosystem survive the ever-increasing levels of human technological and economic activity? The planetary ecosystem consists of a complex, dynamic, but also sometimes fragile, network of interactions, some of which can be disrupted or even irreversibly damaged by human activities.

In recent years, following the report on *Our Common Future* by the Bruntland Commission on Environment and Development in 1987, and the UN Conference on Environment and Development held in Rio de Janeiro in 1992, the term **sustainable development** has come into widespread use to reflect these concerns. The Bruntland Commission defined it in human terms, as 'development that meets the needs of the present generation without compromising the ability of future generations to meet their own needs' (Bruntland 1987: 43).

In this formulation, the emphasis is mainly on material levels of resource use and on pollution, but the term 'needs' is also wider, and might be thought of as also reflecting concerns about lifestyle and quality of life, as well perhaps as global inequalities and redistribution issues.

Some critics of current patterns of energy and resource use go beyond the issue of environmental impacts, resource scarcity and ecosystem disruption. For some, it is not just a matter of pollution or global warming but also a matter of how human beings live. For them, as well as being environmentally unsustainable, modern industrial technology underpins an unwholesome and unethical approach to life. Technology, at least in the service of modern industrial society, leads, they say, not to social progress but to social divisions, conflicts and alienation, and underpins a rapacious, consumerist society in which materialism dominates. Some therefore call for sustainable alternatives to consumerist society (Trainer 1995). Other critics go even further and, from a radical political perspective, challenge the whole industrial project, and the basic concept of 'development', which they see as, in practice, reinforcing inequalities, marginalising the poor, exploiting the weak and disadvantaging minorities, destroying whole cultures and species, and irreversibly disrupting the ecosystem, all in the name of economic growth for a few.

Even leaving aside views like this, from a number of perspectives the interaction between technology, the environment and society would seem to be a troubled one. Clearly it is not possible, in just one book, to try to explore, much less resolve, all these issues. Nevertheless, it may be possible to get a feel for some of the key issues and factors involved.

A model of interactions

What follows is a very simplified model of the conflicting interests that exist in society which may help our discussion in that they may influence interaction between humanity and the environment. Put very simply, there would seem to be three main human 'domains' which interact on this planet with each other and with the rest of the natural environment. First there are the **producers** – those engaged in using technology to make things or provide services. Second there are the **consumers** – those who use the products or services. Third there are those who own, control and make money from the process of production and consumption, chiefly these days, **shareowners**. In addition you might add a fourth meta-group, governments, nationally and supranationally which, to some extent, 'hold the ring', i.e. they seek to control the activities of the other human groups, for example, by developing rules, regulations and legislation.

Obviously these are not exclusive groups: in reality people have multiple roles. Producers also consume, even if all consumers do not produce. Producers and consumers may also share in the economic benefits of the production–consumption process, e.g. as shareholders. Moreover, there will be differences and conflicts within each group: not all the producers, consumers or shareholders will necessarily have the same vested interests. However, in general terms, the conflicts between and among the three groups or roles will probably be larger than those within each group.

Historically, there has been a conflict between producers and owners – labour versus capital, if you like. While the interests of 'capitalists' (i.e. owners or shareholders) are to get more work for less money, battles have been fought by trade unionists to squeeze out more pay. A similar but less politically charged conflict also exists between consumers and capitalists/shareowners. Consumers want good cheap products and services and capitalist/shareowners want profits and dividends. Over the years governments have intervened to control some aspects of both these interactions, for example, in order to limit the health and safety risks faced by workers and to ensure that certain quality standards are maintained in terms of consumer products and services. In parallel consumers have organised to protect their interests.

In some circumstances consumers' and producers' interests may also clash: consumers want good, safe, cheap products and producers want reasonable pay and job security. The 'capitalists' (i.e. the owners and their managerial representatives) tend to have the advantage in most situations: they can set the terms of the conflicts. Thus they may argue that pay rises will, in a competitive consumer market, lead to price increases, reduced economic performance and therefore to job losses. In general they can set the terms of employment and of trade. However, they can be constrained by effective trade unions or by market trends created collectively by consumers.

The environmental part of interaction

So much for the human side of the model. The other element is the natural environment: the source of resources from which producers can make goods for consumers and profits for capitalists. The natural environment has no way of responding actively to the human actors, unless one subscribes to the simplified version of the Gaia hypothesis, as originally developed by James Lovelock, which suggests that the planet as a whole has an organic ability to act to protect itself; the various elements of the ecosystem act together to ensure overall ecosystem survival (Lovelock 1979). Nevertheless, even if the natural environment is passive, it represents a constraint on human activities.

Describing this situation more than a century ago, the German philosopher Karl Marx argued that there could in principle be a conflict between the human actors in the system and the natural environment, but that the constraints on resource availability, and the environmental limits on getting access to resources, were far off. He and his followers therefore devoted themselves to the other more immediate conflicts – between the human actors. Nevertheless, it was recognised that at some point, as human economic systems expanded, they would come into serious conflict with the environment.

You could say that this point has now been reached. The rate of economic growth and technological development has brought industrial society to its environmental limits. Some radical critics argue that such is the desire for continued profit by capitalists that, in the face of effective trade unions on one hand and tight consumer markets on the other, there has been a tendency to increase the rate of exploitation of nature, thus heightening the environmental crisis (O'Connor 1991).

Certainly, mankind has always exploited nature, just as capitalists have exploited producers and consumers, and this process does seem to have increased. However, just as the latter two human groups have fought back, so now the planet is beginning, as it were, to retaliate, by throwing up major environmental problems. As already suggested, apart from putting constraints on some human activities, the natural environment cannot fight back very actively or positively: it requires the help of human actors, environmental pressure groups and governments, to protect and promote what they see as its interests.

The model reviewed

The model is now complete: there are the three conflicting human groups, (producers, consumers and investors/shareholders), locked into economic conflict; governments active nationally and globally to varying extents; and the natural environment. The environment is mainly dependent for protection on the interventions of people and governments, but perhaps it is also able to constrain human activities by, as it were, imposing costs on human activities if these disturb key natural processes and, in the extreme, making human life on earth unviable.

The issues facing those involved with trying to diffuse this complex situation are many and varied; by focusing on economic conflicts, our 'interest' model only partly reflects the political, cultural and ideological complexity of society. In reality people occupy a range of roles: they are not just consumers, producers or shareholders. Moreover, while our model may reflect some of the economic conflicts between various groups within any one country, it does not reflect the conflicts among nations or groups of nations. For example, there are the massive imbalances in wealth and resources among the various peoples around the world and conflicts over ideas about how these imbalances and the inequalities should be dealt with. Moreover, the global nature of industrialisation means that there are global level problems and conflicts, not least since pollution is no respecter of national sovereignty and national boundaries. The world's environmental problems cannot be addressed without considering these wider issues, and our model only partly reflects them.

However, despite these failings, the model does at least provide a framework for discussing some of the key human–environment conflicts. The model abstracts the human element for purpose of analysis, but it should be clear that human beings are not in reality separate from the natural environment. Albert Einstein once said the environment was 'everything except me', but in this book the term **environment** will mean the entire planetary ecosystem and all that exists in it, including human beings. This definition, with humans as part of the environment, is important, in that too often the environment is seen as something outside of humanity – just as a context for human action. The fact that human beings are part of nature makes the problem of solving environmental problems even harder: can the part understand the whole? Will mankind's much vaunted intellectual capacity enable it to rise to the challenge? Or will nature impose its constraints on mankind?

Negotiating conflicts of interest

We shall be returning to some of these questions in Part 4 of this book, but for the moment it seems clear that, assuming that human beings can act usefully to protect the environment, there will be a need to find some way in which the conflicting interests of the four main 'domains' in the model can be balanced. The battles among the three human elements can no longer be allowed to dominate the political and planetary scene: the fourth element, the environment, must also be considered. Indeed some of the purists in the environmental movement would go further: rejecting the idea that minor 'pale green' adjustments and accommodations will suffice, the 'deep greens' would argue, adopting a fundamentalist view, that the interests of the natural world should dominate all others, even to the extent of seriously limiting human activities. This view goes well beyond the idea that human beings have an ethical or moral responsibility for environmental stewardship, which is seen as patriarchal or paternalistic. 'Deep ecology' writers like Devall argue that human beings should stop thinking of themselves as being the centre of creation and instead adopt an 'eco-centric' viewpoint (Devall 1988).

Some of the more pessimistic 'deep greens' seem to believe that the planet will never be safe until the impact of human beings on the environment has been returned to the level it was before human civilisation, or at least industrialisation, got going on a significant scale. At times it almost seems as if the ultimate 'deep green' or 'deep ecology' prognosis is that the planet would only be safe without a human presence and that if mankind does not mend its ways, Gaia will arrange just that.

Social equity

Even leaving aside such ultra-pessimistic views, on the assumption that humankind can respond in time and effectively (a rather a big assumption of course), some 'doom' scenarios still hold force. For example, it seems possible that some responses to environmental problems could involve major social dislocations. Those in control might feel it necessary, in order to protect their own interests, to impose socially inequitable solutions, seriously disadvantaging some specific human group or groups.

If this is to be avoided, then some way must be found to combine environmental sustainability and social equity. There will be a need for some sort of accommodation or balance that will not only ensure a more sustainable relationship between humanity and the rest of the ecosystem, but also attempts to reduce rather than increase social and economic inequalities.

The implication would seem to be that no one human group should be expected to meet all the costs of environmental protection: the costs must be shared. For example, let us assume that consumers want greener products. Industrialists will reply that they can be made available but will cost more. Similarly workers may press their employers for safer and cleaner production technologies both for their own sake and for the communities in which they live. They are likely to be told by industrial managers that this would add cost and could lead to lower wages or even job losses. What is missing in this formulation is the interests of shareholders. After all, you could argue that everyone should carry the burden: consumers, producers and shareholders.

In reality of course, shifting to greener products and cleaner production processes may not in the end cost more, or at least may not be a bad move commercially, in the longer term. As pressures for a 'clean up' grow, for example, through government legislation on environmental protection or consumer pressure for greener products, companies that take the initiative will have a competitive advantage over those that do not. Of course, if left just to market competition, this means there will be some commercial losers and overall there may well be short-term costs and dislocations in terms of employment and profits. That is precisely why a proper negotiation process is important – to reach some sort of agreement on the distribution of costs and benefits among all the human stakeholders, in the wider context of overall environmental protection.

Clearly it will not be easy to establish this sort of negotiation process, even if we stay within the confines of our model and apply it just to one country. Quite apart from the powerful vested interests of the competing groups, regulations take time to have an impact and there are usually ways of avoiding pressures for change, at least in the short term. Once we move the focus beyond a single country, the opportunities for evasion become even greater. For example, companies can move to countries where regulatory pressures are weak. In order to be effective, an attempt has to be made to extend the framework of negotiation and regulation to all levels: local, national and global.

This process of negotiation of interests, at whatever level, will depend to some extent on actions taken by governments by setting new environmental standards and regulations, although, equally, all the human actors in the system can also play a part by including environmental considerations in their otherwise partisan negotiations. Despite the difficulties, there have already been attempts to do this, for example in relation to technological development choices, at all levels around the world.

We shall be looking at examples, and at the concept of social negotiation generally, in Part 4, after we have explored some of the world's environmental problems in more detail and looked at potential technological solutions and their limitations. As will become apparent, purely technical solutions may not suffice if the aim is to develop a genuinely sustainable and equitable future: social and economic changes and adjustments may also be necessary.

The growth of environmental concern

The social and political dimensions of the problem of devising a sustainable future may become clearer if we look back to the beginnings of the contemporary environmental debate. In the late 1960s and early 1970s there was a perhaps unique concurrence of ideas from a number of social and political movements.

The early 1960s had seen the beginnings of environmental concern, symbolised by the publication in 1962 of Rachel Carson's *Silent Spring* which, among other things, warned of the ecological dangers of pesticides such as DDT.

Subsequently there was a growth in environmental concern among young people, many of whom formed part of a counter-culture which flowered briefly in the late 1960s and early 1970s. They were often from relatively affluent backgrounds, but they challenged the ideas of the conventional consumerist and materialist society in which they had grown up. Some were not content with simply objecting to the way things were done at present but also wanted to create alternatives: alternative lifestyles and alternative technologies to support them. There were self-help experiments in rural retreats with windmills, solar collectors, and so on.

In parallel, the late 1960s and early 1970s saw a rise in radical politics, reflected most visibly by the student protest movements around the world. The 'new left' that emerged as one of the many strands in this movement, challenged the political dogma of the traditional left. The latter held that capitalism, with its single-minded concern for the economic interests of those who owned and controlled technology, denied society as a whole the full benefits of technology; but once freed from capitalist constraints the same technology could be used to meet human needs more effectively. The classic interpretation of this view in practice had been fifty years earlier when, following the Russian Revolution, Lenin adopted Western production technology, since he argued that it was the best available at that historical stage. The theoretical point was that the existing technology could simply be redirected to meet new ends. Eighty years on, the shortcomings of this view have become clear: industrial development during the Soviet period seems to have replicated many of the worst examples of environmentally destructive Western technology.

The new left in the late 1960s and early 1970s to some extent foresaw this problem. They argued that technical means and political ends inevitably interacted: old means could not be used to attain the new ends, a new set of technologies was required. Like many environmentalists and the 'alternativists' in the counter-culture, they were also arguing for an alternative technology.

Alternative technology

This line of argument was put forward by a British writer, David Dickson, in *Alternative Technology: The Politics of Technical Change* (1974). However, Dickson, along with many members of the counter-culture, also felt that a simple switch of technology would not be sufficient. Technology and society interacted, so there was a need for an alternative society as a base for the alternative technology: 'A genuine alternative technology can only be developed – at least on any significant scale – within the framework of an alternative society' (Dickson 1974: 13).

Not all the environmentalists agreed: some felt that reforms would be sufficient, and that technology was in any case a separate 'technical' issue. Others adopted a more strategic approach. For example, Peter Harper, an English enthusiast for what he called 'radical technology', claimed that: 'Premature attempts to create alternative social, economic and technical organisation for production can contribute in a significant way to the achievement of political conditions that will finally allow them to be fully implemented' (Harper 1974: 36).

Others again warned that any attempts to introduce radical alternatives would be co-opted by commercial interests and would be shorn of their radical political edge. The emphasis could be on selling 'technical fixes', that is just the hardware, and not on implementing the social changes with which alternative technology was meant to be associated. Thus, for example, solar collectors could become just conventional consumer products rather than harbingers of social transformation.

The current 'green' debate

These debates are relevant to our contemporary situation in that the issues are now much clearer. Some alternative technology has been co-opted, some technical fixes are being offered as solutions to our environmental problems, while at the same time some radicals in the contemporary 'green' movement, are still arguing that only a radical transformation of society will be sufficient.

Inevitably, there is a wide range of views on these issues in the 'green' movement. This is hardly surprising since the green movement, which emerged in the 1980s, has a multitude of strands. It consists of much more than just the members of green political parties or activist groups, even though some of these are now quite large: tens of thousands of people belong to organisations such as Friends of the Earth and Greenpeace. In addition, it might also be thought to cover anyone who has some concern for the environment, as reflected, for example, in their support for wildlife protection or in their consumer behaviour.

All these levels of involvement can have an impact, although equally there can also be tactical and strategic divergences and disagreements. Certainly, the growth of consumer awareness has led to pressure for environment friendly products and, in turn, for environmentally sound production processes: greener products and cleaner production technologies which have fewer impacts and use less energy. There is considerable activity in this field at present, but equally there are those who would ask whether this is enough to achieve real sustainability. For example, it has been argued (by Chris Ryan, a leading Australian exponent of ecodesign) that if the various global environmental problems are to be properly addressed, pollution levels and global energy and material resource use must be cut by around 95 per cent, but that this may not be possible just by 'technical fixes'. There may also be a need for social change, for example, in qualitative patterns and quantitative levels of consumption (Ryan 1994).

For radical 'greens' the real issue is: can and should growth in material and energy consumption be continued, stimulated by ever-growing expectations concerning living standards? Is there not a need for more radical changes – in society as well as technology? Increasingly it is argued that there is a necessity for a more radical transformation of technology and also possibly of society – an alternative set of technologies better matched to environmental protection, linked to an alternative set of social and cultural perspectives and structures.

Rather than explore these large issues in the abstract, this book attempts to look at the case for and the implications of this type of change by focusing on energy technology as an example. Clearly this is just one area of technology but, as has been indicated, it is perhaps the central one in environmental terms. The next chapter provides a basic introduction to energy issues, reviewing energy use past, present and future.

Summary points

- Energy use has increased dramatically in recent years and this has had increasingly adverse impacts on the environment.
- The concept of sustainable development has been proposed, which implies that alternative technologies may be needed which avoid or reduce adverse impacts.
- Although alternative technologies may help, there may also be a need for major social changes if sustainability is to be achieved.
- All the various groups in society must share the burden of protecting the ecosystem of which they are all inevitably part.
- Energy use is probably one of the key factors influencing environmental impacts.

Further Reading

A useful general overview of the way technology has shaped human–environmental interactions is provided by David Kemp's *Global Environmental Issues* (1994, Routledge, London). Some of the underlying philosophical issues discussed in this chapter are explored in some depth by Adrian Atkinson in *Principles of Political Ecology* (1991, Belhaven, London). For a good introduction to the history of green thinking see David Pepper's

Modern Environmentalism (1984, Croom Helm, London) and Andy Dobson's *Green Political Thought* (1995, Routledge, London).

You might also like to try to track down David Dickson's classic text *Alternative Technology: The Politics of Technical Change* (1974, Fontana, London). Although this is out of print, it should still be available in libraries. The various strands of thinking which make up the 'deep green' or 'deep ecology' approach are explored in the classic text by Bill Devall and George Sessions *Deep Ecology* (1985, Peregrine Smith Books, Salt Lake City). The latest book by George Sessions, *Deep Ecology for the Twenty-First Century*, is available from Schumacher Books, Foxhole, Dartington TQ9 6EB, UK, along with several other similar titles.

For a somewhat more conventional viewpoint on environmental problems and policies see *Our Common Future* by the Bruntland Commission on Environment and Development (1987, Oxford University Press, Oxford). For a useful overview see Jonathon Porritt's *Where on Earth are we going?* (1990, BBC, London).

2 Energy and environment

- **Energy units**
- **Acid rain**
- **Global warming**
- **Nuclear opposition**

In order to explore the environmental implications of energy generation and use, we will need an understanding of the basic terms, concepts and measurement units used in the energy field. As you will discover, perhaps surprisingly, some of the measurement units are far from uncontroversial. Having established this context, the chapter then looks at how energy is and has been used around the world, and at some of the social and environmental problems that have emerged as a consequence of the use of fossil and nuclear fuels. Finally, we look briefly at what the alternative energy options might be.

Energy and its use

Energy is a concept rather than an actual thing: we say people have energy when they can work or play hard. The manifestation of energy in material terms is 'fuel', and these two terms, energy and fuel, tend to be used interchangeably. The concept of power is also often used as if it meant the same as energy.

To set the scene we need to have a clearer understanding of some of the basic units and terms used in the discussion of energy issues.

Although it is common to talk of 'energy generation' and 'energy consumption', strictly, energy is never 'created' or 'consumed', it is just 'converted' from one form to another. The term **power** is used to describe the conversion capacity of any specific device, i.e. the rate at which it can convert energy from one form to another, and the unit most commonly used is the **watt**. Strictly it is a measure of the 'capacity to do work'.

Specific energy generating or consuming devices are therefore given a **power rating** (or **rated capacity**) in watts and multiples of watts, e.g. a **kilowatt** (kW) is 1,000 watts. The **megawatt** (MW) is 1,000 kilowatts (or 10^6 watts), the **gigawatt** (GW) is 1,000 MW (or 10^9 watts), and the **terawatt** (TW) is 1,000 GW (or 10^{12} watts). To give you an idea of scale, a typical large modern coal or nuclear power station has a rated capacity of around 1.3 gigawatts (GW), while in the mid-1990s the UK had around a total of 65 GW of electricity generating capacity.

The amount of **energy** converted (generated or consumed) or more accurately, the actual 'work' done, is defined by the power of the device multiplied by the time for which it is used (i.e. watts × hours). It is usually measured in **kilowatt hours** or (kWh). This is the unit by which electricity and gas is sold in many countries (although, obscurely, the USA still makes use of British Thermal Units, the old measure for the heat content of fuels: 1 kWh = 3,413 BTUs). A typical 1kW rated one-bar domestic electric fire consumes 1 kilowatt hour (kWh) each hour.

For larger quantities, multiples of kWhs are used, most commonly the **terawatt hour** (TWh) which is 1,000,000,000 kWh or 10^9 kWh. To give an idea of scale, the national mid-1990s figure for total UK electricity consumption was about 300 TWh per annum. Remember, however, that this is the figure for the consumption of electricity, not total energy consumption: it does not include all the direct **heat** supplies (gas, etc.) or **transport** fuels.

The total amount of energy used is often measured in terms of **primary energy** consumption, that is the amount of energy in the basic fuels used by energy conversion devices, whether for electricity production, heating or transport.

However it is important to remember that primary energy figures, for the total energy in the fuels used by energy conversion devices, are much larger than the finally delivered energy, as utilised by consumers, since there are losses in the conversion process in power plants and in transmission along the grid cable network. This is particularly true of electricity: conventional coal or nuclear fired power plants only have conversion efficiencies of around 35 per cent. Even the best modern gas-fired power stations can only convert around 50 per cent of the energy in the input fuel into electricity. Moreover, after it has been produced, up to 10 per cent of the electricity may be lost when it is transmitted along power lines to consumers, depending on the distances involved. Finally, consumers will use this 'delivered' energy to power a variety of energy conversion devices with varying degrees of efficiency, much of it often being wasted, for example, in poorly insulated buildings. Primary energy figures therefore only tell part of the story. As we shall see in subsequent chapters, there is also a need, when comparing technologies and energy systems, to consider the overall efficiency of energy conversion and transmission, and the use to which the energy is put.

The battle of the units

As can be seen from the previous discussion, measuring energy use is not as simple as it might seem. Given that there are many ways in which energy is generated and used, it is not surprising that there are many different, often confusing, ways in which it is measured and many devotees of rival systems of measurement. We have mentioned kWh, which is the most familiar unit to most people since it is what is used on consumers' bills. However, energy analysts sometimes use the basic physical unit for 'work', the **joule** (J) or multiples of joules. One watt is one joule per second, so a kWh is 3,600,000 joules, and the joule is thus a very small unit. Hence large multiples are common, e.g. peta-joules or PJ (1,000 tera-joules) and exa-joules or EJ (1,000 peta-joules).

In the UK until recently, for statistical comparison purposes, primary energy use was also often measured in terms of the equivalent amount of coal that would be required to be burnt to provide that energy regardless of what fuel was actually used in power stations, i.e. in 'tonnes of coal equivalent' (or more usually 'million tonnes of coal equivalent' or 'mtce'). This no doubt reflected the historical predominance of coal in the UK's economy, although, to confuse matters, the therm was also used. In 1994, to make the situation perhaps a little clearer, the UK government's statisticians decided to adopt the European standard unit, with the energy content of all fuels being rendered, for statistical comparison purposes, in terms of the equivalent amount of **oil** that would have the same amount of energy content. The energy content of all fuels is therefore now presented in terms of **tonnes of oil equivalent**, or more usually, **million tonnes of oil equivalent** (mtoes).

However, for the purposes of this book we will stay with the hopefully more familiar kWh, TWh, etc. figures. For reference 1 mtoe = 11.63 TWh, and 1 TWh = 0.086 mtoe.

Not everyone will find the details of the units used in energy measurement exciting, but obviously it is necessary to have a common measure, and establishing this may not always be as uncontentious as it may seem. In this regard, before we move on to look at the way energy is actually used, it may be worth noting an extra complication concerning how it is measured.

In addition to switching to the use millions tonnes of oil equivalent, the changes introduced in 1994 to the way energy statistics are rendered in the UK also involved a subtle shift in the way the energy content is calculated. It is now based on the energy content of the output power produced by plants, rather than on the energy content of the input fuel needed to generate the power. This has some interesting results. Previously, on the so-called fuel substitution basis, the energy contribution from non-fossil fuel powered devices like the wind turbines was calculated in terms of the energy content of the fossil fuel that would have to be fed to a conventional power station to provide the same amount of power output. On the new 'energy supply' basis, the figures for contributions from devices like wind turbines drop dramatically, by 73 per cent, this being the scale of the losses associated with energy conversion in conventional power plants. The end result of the change was that the contribution in 1993 from the so-called renewable energy sources (hydroelectricity, wind power, etc.) became 0.4 mtoe instead of 1.4 mtoe (Department of Trade and Industry 1994a).

Not surprisingly, the new approach is not particularly popular with renewable energy supporters. Interestingly the figures for nuclear power plants are not much affected by this new way of rendering energy statistics. Fortunately though, the figures are also made available on the original basis, since it was accepted that while it was useful to be able to compare the actual amount of power being supplied to the grid, it was also useful to be able to assess the degree to which renewable sources were substituting for fossil fuel. But this methodological aside should at least alert you to the need to be careful about the way units are used.

National and global energy use

Having now established some of the basic energy units, we can move on to look briefly at how energy is actually used. Primary energy figures can be derived at various level – for countries, or for the world as a whole. Within the national context, primary energy use is often broken down in terms of its final destination, i.e. in terms of its eventual end use in each sector of the economy.

Inevitably the exact figures change with time, but to give an impression of the balance among the various main end use sectors, in the UK in 1994, **transport** accounted for 33 per cent of primary energy use, **industry** 25 per cent, the **domestic sector** 29 per cent, leaving 13 per cent for other uses. In terms of the actual form of energy consumed, in the UK in 1994, electricity accounted for 16 per cent of the total, gas 32 per cent, petroleum 44 per cent and solid fuels 8 per cent (Department of Trade and Industry 1995).

Energy conservation techniques are beginning to have an impact in some sectors, leading to overall reductions in energy use in some cases. For example, the UK has managed to achieve around a 40 per cent cut in energy use in the industrial sector since 1970. However, energy use in the domestic and transport sectors has stayed more or less constant as has overall energy use in the UK, despite the advent of more energy efficient technologies (Department of the Environment 1996).

In some other sectors there has been an increase with, interestingly, the use of computers having an impact on energy consumption. For example, in the USA the use of computers currently accounts for around 5 per cent of the commercial electricity load and it is the fastest growing segment, expected to reach 10 per cent by the year 2000. New more energy efficient computers are being introduced, and computer use should in principle increasingly save energy, for example, by reducing the need for some business travel and by improving the efficiency of business and commercial activities generally. Even so, it is worth noting that, worldwide, computers currently use around 240 TWh of electricity each year, not far short of the UK's total power usage (Young 1993).

The pattern of energy production has also undergone some significant changes in recent years, particularly in relation to electricity generation. Thus in 1992, 60 per cent of the UK's electricity came from coal-fired power stations, 21 per cent from nuclear plants, 8 per cent from oil-fired plants, 2 per cent from hydroelectric plants and 4 per cent from gas and other fuels, with 5 per cent being imported (Department of Trade and Industry 1993). But by 1995 only 47 per cent of the UK's electricity came from burning coal, while gas had a projected 16 per cent share, primarily due to the use of natural gas fuelled turbines for electricity production (Department of Trade and Industry 1996).

Similar trends are occurring elsewhere in the world, with gas becoming a major new source of electricity as well as heat. Table 2.1 shows the global primary energy use in 1992, by source. Whereas the pattern of supply and demand within individual countries can, as we have seen, vary significantly from year to year, the global percentages, which

Table 2.1 ***Global primary energy consumption in 1992, by source (%)***

Hydro	5.9
Nuclear	5.6
Oil	33
Biomass	13.8
Gas	18.8
Coal	22.8

Source: British Petroleum, *BP Statistical Review of World Energy*, BP Corporate Communications Service, 1993.

aggregate the patterns of a large number of countries, do not tend to change rapidly: for example, the 1994 data for global energy consumption show a very similar pattern of energy consumption and fuel use. However, the overall trend globally is upwards; the 1995 figures showed an overall 1.8 per cent rise in world energy use.

Looking further into the future, given the growing world population, energy use globally is likely to continue to increase. Later in this book, we will be looking at a range of energy scenarios which have been devised to attempt to map out possible patterns of longer term development in energy supply and demand. Some assume that energy demand is likely to increase up to perhaps three times current levels by the year 2060.

However, predicting the future is difficult. Past trends are not necessarily a good guide to what might happen next in terms of fuel availability and prices, or in terms of patterns of energy consumption in the various rapidly changing sectors of energy use around the world. The future is far from being predetermined: part of the aim of this book is to explore what choices there are in terms of creating new patterns of supply and demand.

The growth of energy use

Having now set the statistical scene in terms of general patterns of energy use, let us now look in more detail at how this pattern emerged historically. For, although past trends may be a poor guide to future patterns of development, in order to begin to think about the future, we need to look at how the present pattern of energy use came about.

In ancient times and right up to the beginning of the industrial revolution, motive power was provided either by direct human effort, including the use of slaves, by animals, or at sea, by wind. Natural fuels such as wood were used for heating, cooking and some processing of materials. In the Middle Ages machines such as windmills and watermills gradually took on some of the load; watermills actually played a key role in the early stages of the industrial revolution. However, the industrial revolution only really got going when coal began to be used widely in the mid-eighteenth century. Whereas before factories had to be sited near power sources such as rivers and also possibly near sources of raw materials, coal could now be transported to factories – in trains powered by coal-fuelled steam engines – to run the factory equipment. Coal also provided a fuel for trains and ships to shift raw materials from remote sites to factories, and to transport their finished products to distant markets. Moreover, coal-fired engines could be used to pump water out of mines, thus allowing deeper mines to be used and more coal to be produced. The combination of coal, steam engine and train technology was thus a real industrial and economic breakthrough.

The role of coal as the key fuel was not challenged for hundreds of years: coal was used for heating houses and was also converted into gas for use in lighting and heating; it powered trains and was fed to power stations to generate electricity. In some countries use could also be made of hydroelectricity from large dams, but in general the first main rival to coal was petrol – petroleum spirit extracted from oil – which found a use in some industrial processes and increasingly as a fuel for vehicles. Initially, in the 1890s, most of the first automobiles were powered by electricity or steam, produced using coal, but by the beginning of the twentieth century the petrol-powered internal combustion engine had taken over.

In the inter-war years, while electricity became increasingly important in many sectors of the economy, oil and gas also gradually took on increasing roles. In the years after the Second World War, oil began a steady rise and by the time of the first oil crisis in 1973–4, its use, at least in the advanced industrialised countries, had overtaken coal. At this point a new direct source of natural gas was also found – from under the seabed – and natural gas rapidly replaced the so-called ‘town gas’ that had previously been produced from coal. In parallel, following the war effort to produce an atomic bomb, civil nuclear power had begun to make a small contribution to electricity supplies in most of the developed countries.

So as the modern industrialisation process got underway in the advanced Western countries in the period after the Second World War, the basic energy pattern was fairly stable: coal, oil and gas had very roughly equal shares in primary energy use terms, but with oil beginning to dominate, and with nuclear power (along with hydroelectric plants) making a small additional contribution. Basically, coal provided the bulk of electricity, gas provided the bulk of heating, oil the bulk of transport fuel. The exact proportions differed from country to country, depending on the availability of indigenous reserves. For example, the UK had ample coal, as did the USA. In addition the USA had oil, although it also imported oil from the Middle East, as did the UK and many other industrialised countries. By contrast, some of the newly emerging countries like Japan had few coal reserves and no oil and had to rely heavily on imported fuel.

The oil crisis

The 1973–4 oil crisis was precipitated by the Yom Kippur War between Israel and the Arab states. The Organization of Petroleum Exporting Countries (OPEC) which at that point was dominated by the Arab states, objected to the West's support of Israel and imposed a sudden increase in oil prices. This had a dramatic effect on the industrialised economies: they suddenly realised how reliant they had become on imported oil. Over the next decade strenuous technical efforts were made to diversify away from oil, although equally strenuous political, and in some cases military, efforts were also made to secure access to Middle East oil.

The oil crisis coincided with the first serious signs of concern about the environmental impacts of the use of fossil fuel, although the main concern, as highlighted graphically by the oil crisis, was about fuel scarcity. As a result of what was seen as an 'energy crisis', a range of new technologies were looked at – including the use of natural energy sources such as the winds and solar heat, which are collectively called **renewable** sources, since, unlike the fossil fuels, they will never be exhausted. Nuclear power, which had been developing relatively slowly in the West since the war, was also given an extra boost.

By the early 1980s, however, the pattern had settled back down: oil prices had dropped in real terms and natural gas had become a new and cheap fuel. Nuclear power had expanded, but not as dramatically as its proponents would have liked. It was bedevilled by economic and technical problems. The accident at Three Mile Island in Pennsylvania in 1979 effectively halted progress in the USA; worldwide the total nuclear contribution hovered around 5 per cent of primary energy, reaching 5.7 per cent by 1990. Renewable energy technology was still in its infancy, although it is important to remember that hydroelectric dams generate around 20 per cent of the world's electricity from a renewable resource, and in some countries (notably Brazil and Norway) hydro power provides the largest single energy input. Overall though in most industrial countries coal, oil and increasingly gas remained the dominant options.

Environmental problems emerge

Although environmental issues became prominent in the early 1970s, some of the key issues had emerged earlier. For example, environmental pollution from coal burning became a key issue in the UK as a result of a series of disastrous smogs in the 1950s, when there were many deaths. One smog in London in December 1952 lead to an estimated 4,000 deaths in the following weeks. One response was the 1956 Clean Air Act, which limited the use of coal in open grates in urban areas. An attempt was also made to disperse emissions from power plants and industrial chimneys by increasing the heights of the chimney stacks. This technical fix worked up to a point – although the result was that some of the key pollutants were carried over longer distances and began to have an impact far away, for example on the Scandinavian ecosystem.

Similar problems existed with emissions from power plants elsewhere in the world with, for example, Germany's Black Forest also suffering damage. By the 1980s the problem could no longer be ignored. As a result of objections by the Scandinavians and the campaign work of environmental groups, there was a renewal of environmental concern over the impacts of the use of fossil fuel. By 1983 Germany had embarked on a programme of removing sulphur dioxide gas from power station emissions, although some other countries like the UK took longer to react.

The key issue was that of **acid rain** – the consequence of acid emissions from power stations, caused primarily by the sulphur content of coal and oil. When burnt, the small sulphur element in these fossil fuels is converted into sulphur dioxide gas, which can dissolve in water to produce weak sulphuric acid. Precipitated in the rain, this acid can damage trees and, collected up in lakes, can injure wildlife and fish. Acid rain also damages buildings and crops. Additionally, oxides of nitrogen can be produced by the combustion of fossil fuels in power stations and from the combustion of petrol in cars, and they can play a similar role.

In principle, the acid emissions from the combustion of fossil fuels can be filtered out relatively easily – although at a cost. By contrast there was no obvious remedial technology for the next problem that emerged, that is the emission of greenhouse gases such as carbon dioxide. Carbon dioxide (CO_2) is the fundamental product of combustion: put simply burning means using oxygen from the air to convert the carbon in fossil fuels into carbon dioxide gas plus heat.

Fossil fuels are hydrocarbons: they contain varying amounts of hydrogen and carbon. All fossil fuels create carbon dioxide when burnt, but some produce more than others depending on their chemical make-up. Coal has a high pure carbon content, and when burnt it therefore produces mostly just carbon dioxide and heat. By contrast methane (natural gas) is made up of one atom of carbon plus four of hydrogen and when burnt the hydrogen is converted into water, so that the ratio of carbon dioxide to heat produced differs. The result is that the combustion of natural gas produces roughly 40 per cent less carbon dioxide per kWh of heat produced than the combustion of coal. By contrast the combustion of the more complex hydrocarbons like oil, produces intermediate levels of carbon dioxide. See Table 2.2.

Table 2.2 ***Carbon dioxide production***

Fuel	*kg of CO_2 per GJ of heat*
Coal	120
Oil	75
Natural gas	50

Carbon dioxide is used in fizzy drinks: it is a tasteless, colourless and chemically inactive gas which dissolves only very slightly in water. This makes it difficult to remove from power station or car emissions. In theory you could freeze it out from power station emissions, producing solid carbon dioxide or 'dry ice' – at vast expense. Alternatively, you could collect and store the gas, again at vast expense. However, in practice the only realistic way to avoid carbon dioxide emissions is not to burn fossil fuels.

Global warming

Global warming is of course the reason why there is a need to avoid producing carbon dioxide. Gases like carbon dioxide travel up into the upper atmosphere (the troposphere) where they act as a screen to sunlight. They allow the suns rays in but stop the heat radiation from re-emerging, much as happens with the glass in a greenhouse. The result is that the greenhouse, in this case the whole world, heats up. Some degree of global warming is actually vital, otherwise this planet would be too cold to support life. However, the vast tonnage of carbon dioxide gas we have released into the atmosphere seems likely to upset the natural balance.

The situation has been made worse by the fact that large areas of forest around the world have been felled, thus removing an important 'sink' for carbon dioxide, since trees absorb it as they grow. Reafforestation is obviously vital, but it would take many decades just to replace what has been lost and, unless carried out on a vast scale, could only play a very small role in absorbing the vast amount of carbon dioxide released every year by power plants and cars.

Carbon dioxide is not the only culprit. Methane gas, produced naturally from marshes, cows and other ruminating animals and also from human wastes, plays a very significant role in global warming – much more, molecule for molecule, than that played by carbon dioxide. Ozone gas also plays a part in the complex chemical interactions in the troposphere, affecting the way in which the sun's radiation is absorbed.

However, it is worth noting at this point that the greenhouse effect is unrelated to ozone depletion and the creation of so-called 'ozone holes' in the stratosphere. Although ozone plays a role in the greenhouse effect, the ozone holes are a separate phenomenon – the result of a chemical interaction between CFC (chlorofluorocarbon) molecules, which destroy ozone, notably in polar regions. CFCs are manmade chemicals which were developed as supposedly inert gases for such uses as foam packaging or in refrigerators. They are also used as propellant gases in aerosol cans.

The results of stratospheric ozone depletion due to the release of CFCs are that dangerous wavelengths of solar radiation can reach the earth's surface where they can cause cancers and damage plant growth.

The results of global warming, if it happens on a significant scale, are likely to be even more severe. It should be remembered that the ice age only involved a global temperature variation of around 4 °C. Global warming could result in the icecaps melting and this, coupled with the effects of the thermal expansion of the seas, would cause sea levels to rise. Global warming could also lead to the disruption of crop growing as climate patterns change. It would not be simply a matter of increased temperatures: the climate system would become erratic, with more storms and more droughts. Flooding could be very severe in some parts of the world, particularly in low-lying areas such as the Netherlands and Bangladesh, with consequent great risk to life and the loss of areas for habitation and cultivation. Some low-lying islands, for example in the Pacific, could disappear entirely.

Given that the impact of global warming on life on earth could be very dramatic, insurance companies around the world are already taking the issue seriously. Some even claim that the effects have already started. Most scientists are not yet prepared to pronounce finally on this, although the consensus is that global warming is a strong likelihood. For example, in its 1995 assessment, the Intergovernmental Panel on

Climate Change (IPCC) suggested that, on the basis of the most up-to-date models, average global temperatures were likely to rise by between 1 and 3.5 °C by the year 2100, with 2 °C being their best estimate. This might lead to average sea level rises of between 15 cm and 95 cm by 2100, the best estimate being 50 cm (IPCC 1995).

However, there are disagreements about exactly what is happening and what is likely to happen, and much general speculation. The problem is that it will probably be some decades before the likely extent of global warming is known for sure, and by then it may be too late to respond effectively, except, where possible, by building dykes. Indeed this might not even be a viable response if some of the more extreme predictions prove to be correct: for example, it is conceivable that when and if temperature begins to rise above a certain point, an ever more rapidly accelerating 'runaway' greenhouse effect could occur, as the warming seas are unable to absorb carbon dioxide, and more methane is produced from the spreading wetlands (Legget 1991).

Political reactions

In the 1990s the threat of global warming and climate change has been taken increasingly seriously by governments around the world, although not all are as yet prepared to adopt the 'precautionary principle', which would suggest that action should be taken now despite the absence of full scientific identification of the scale of the problem. However, some governments are willing in principle to adopt what is called the 'policy of least regret', that is to support developments which would be sensible even if global warming does not turn out to be such a significant problem. The result is that relatively modest measures have been adopted: the Rio Earth Summit in 1992 produced an agreement backed by more that 160 countries to try to get carbon dioxide emissions back to 1990 levels by the year 2000. Unfortunately, not all of the signatories seem likely to achieve even this relatively moderate stabilisation target, and key countries such as the USA and some Arab oil states have been less than enthusiastic about the idea. To put the situation in context, the IPPC has suggested that net reductions in emissions of up to 60 per cent below 1990 levels would be required in order to stabilise climate change.

More radically still, a German parliamentary commission and a report to the Netherlands Ministry of the Environment both found that to limit the global temperature rise to about 2 °C, industrial countries like Germany and the UK would have to cut their CO_2 emissions by 75 per cent by

2030 and 85 to 90 per cent by the year 2050. This reduction reflects estimates of Third World population growth and basic development needs and assumes that the main burden of reducing CO_2 falls on rich countries, which have contributed most of the carbon to date (Deutsche Bundestag 1991; Krause *et al.* 1989).

Clearly there is some way to go before the world community can think in these terms. At present the most ambitious proposal, from the coalition of low lying island states most threatened by any potential sea level rises, is for a 20 per cent reduction by 2005, and this has not yet won widespread support.

Technical alternatives to fossil fuels

There has however been some enthusiasm for a return to the idea of using nuclear power – since nuclear plants do not generate any carbon dioxide gas. As we shall be discussing later, the scale of nuclear expansion that would make a significant impact would have to be very large, and it seems unlikely that this will happen. Although some countries, like France, Japan, China and Korea, remain keen on nuclear expansion, in most of the rest of the world nuclear power is perceived as a failed or at least stalled option. In part the problem has been economic. The 'energy crisis' envisaged in the mid-1970s failed to materialise: fossil fuel prices subsequently fell and nuclear power could not compete. Nuclear expansion had also become hard to support politically in many countries due to public opposition.

As Box 1 illustrates, the 1970s saw the beginning of objections to nuclear power from environmental groups, based primarily on safety concerns. In some people's view, these were confirmed by a spate of nuclear accidents – for example, at Three Mile Island in the USA and Chernobyl in the Ukraine. Popular opposition throughout the industrial world, some of it very militant, led to nuclear programmes being abandoned or slowed and the decision by some countries such as Denmark not even to attempt to go down the nuclear route.

As Figure 2.1 illustrates, following the Three Mile Island accident in the USA in 1979, support for nuclear power in the UK dropped notably, and it fell even further after the Chernobyl accident in 1986. Subsequently, after a slight revival, it has tailed off dramatically, while opposition has continued to grow. By 1991 78 per cent of people in the UK interviewed by Gallup either wanted 'no more nuclear plants at present' or for the use of nuclear power to be halted.

Box 1

Nuclear opposition

It is worth remembering just how strong opposition to nuclear power has been. It became most visible when nuclear plant construction programmes began to expand around the world following the 1973–4 oil crisis. In the USA there were mass demonstrations at nuclear sites, for example, in May 1977 at the site of the proposed reactor at Seabrook in New Hampshire where 1,400 people were arrested. These actions were non-violent, but in Europe larger and more militant demonstrations took place: for example, in November 1976 30,000 people attended what was planned to be a peaceful demonstration against the planned reactor at Brockdorf in Germany. Some 3,000 tried to occupy the site and there were violent clashes with the police who used water cannons, tear gas grenades and baton charges to try to restore order. Similar battles took place at Grohnde near Hamelin in March 1977. In July 1977, during a major demonstration against the French prototype fast breeder reactor at Malville, involving more than 60,000 people, one demonstrator was killed.

The demonstrations nevertheless continued: in September 1977 60,000 people protested at the site of the proposed fast breeder reactor at Kalkar in Germany and in the same month in Spain 100,000 people joined a protest in Saragossa, while 600,000 took part in a demonstration against plans for a reactor at Lemoniz (Elliott 1978).

Of course some of these protests might be written off as the results of anti-establishment agitation among a volatile student movement. However, opposition to nuclear power spread well beyond the fringe. For example, in the 1980s in the UK, when attempts were made to find sites for low level nuclear waste repositories they were met by blockades from local farmers and militant opposition by residents, even in very conservative parts of the country (Blowers and Lowry 1991).

Public opposition on this scale has clearly affected decision-making about nuclear power, with governments being aware of its unpopularity. More generally, environmentalist pressure on safety issues has forced up the price of nuclear power and this, combined with the relatively low level of fossil fuel prices, has had the effect of further undermining the economics of the industry. These trends, coupled with the break-up of state-financed nuclear companies in the UK and the tighter economic climate in the USA, have effectively killed off hopes of future expansion. Overall, according to the proponents of nuclear power, by the year 2000 it might, at best, contribute 7 per cent of the world's primary energy (Maisseu and Delanoe 1995).

Figure 2.1 ***Public opinion on nuclear power in the UK 1979–91***

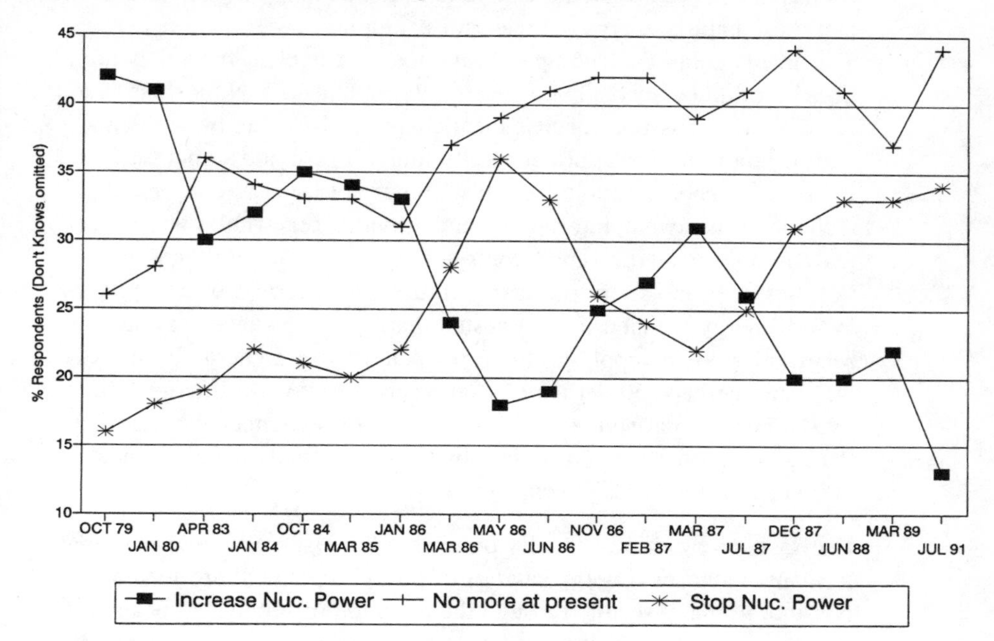

Source: Social surveys (Gallup) from the UK Government's Nuclear Review, Submission by Stop Hinkley Expansion, September 1994.

The longer term prospects for nuclear power are unclear. Fission reactors use a fuel (uranium) which, although still relatively abundant, will not be available indefinitely; fast breeder reactors, which in effect would stretch the availability of the fuel, have yet to be operated commercially and pose what some people feel are significant safety and security risks. Finally, nuclear fusion remains a long-term possibility but, as we shall see later, even if the technology can be perfected, it too has its own problems.

Alternative energy options

Rather than try to create little artificial suns on the earth in the form of fusion reactors, many environmentalists believe it is more immediately credible to make use of the natural fusion energy that the sun already produces and which reaches us as sunlight. Ever since the 1973–4 energy crisis, research on solar power and other forms of **renewable energy** has expanded and led to some relatively large-scale deployments. For example, by 1995 there were around 4.5 gigawatts of wind turbine generating capacity in place around the world.

As has been noted earlier, the term renewable energy is used to indicate that these natural energy sources (for example, sunlight, winds, waves and tides) cannot be used up – unlike fossil or nuclear fuels they are not based on finite reserves but are naturally replenished. Nevertheless, as we will be discussing in subsequent chapters, there can be problems with trying to use what are generally more diffuse and sometimes intermittent energy sources. Even so, renewable energy conversion technology is developing rapidly, and although renewables are currently making only relatively small contributions to overall world energy supplies, they look very promising in the longer term. For example, the World Energy Council has suggested that, given the necessary support, renewables could supply up to 30 per cent of world energy by the year 2020 and perhaps 50 per cent by the year 2100 (World Energy Council 1994). For comparison, a study carried out for Greenpeace suggested that, given proper support, renewables might actually supply almost 100 per cent by 2100 (Greenpeace 1993).

Energy conservation is equally promising but also still rather marginal in many countries. Energy savings of 50 to 70 per cent are seen as possible in many sectors through the adoption of relatively cheap measures; yet the incentive to invest in energy saving is so far generally less than the incentive to invest in new energy supply options.

Clearly in the case of both renewable energy and energy conservation there seems a need for a more forceful approach if the global warming problem really is as significant as many believe.

There have been attempts by some governments to promote the development of renewables and conservation by the use of grant aid, subsidies and other financial incentives. There have also been proposals by the European Commission for carbon taxes which would penalise fossil fuel use and stimulate alternative energy developments. At the same time some renewable energy technologies and energy conservation techniques are beginning to be taken up under the influence of conventional market pressures – since they are commercially attractive in some circumstances. However, not everyone is convinced that the problem of developing a sustainable energy system can be resolved just by throwing money at technology, by imposing broad taxes or by leaving it all to the market. What is needed, it is argued, is an overall strategy to guide energy developments.

Part 2 of this book reviews some of the technical options. To set the scene Chapter 3 attempts to develop some basic criteria for a sustainable energy system as a guide for the future, drawing on the discussion so far. This may provide us with a way of assessing the various energy options.

Summary points

- The combustion of fossil fuels is creating major global environmental problems, most notably global warming.
- Alternative technologies exist which may resolve some of these problems.
- Nuclear power is seen by some as one possibility, but it has economic problems and has met with widespread public opposition based on concerns about safety.
- Although renewable energy technology has its own problems, if combined with energy conservation, it may be a better solution.
- We need to have some criteria to help us assess and choose among the various energy options.

Further reading

The Open University Environment course reader *Energy, Resources and Environment*, J. Blunden and A. Reddish (eds), co-published by Hodder and Stoughton in 1991 and revised in 1996, contains an excellent introduction to energy concepts and energy-related environmental problems. Janet Ramage's *Energy – A Guidebook* (1997, Oxford University Press, Oxford) is one of the most useful general introductions to energy studies.

For an interesting history of energy use in the world over the centuries see Jean-Claude Debeir, Jean-Paul Deleage and Daniel Hemery, *In the Servitude of Power: Energy and Civilization through the Ages* (1991, Zed Books, London and New Jersey).

There is a vast literature on global warming and climate change, ranging from the technical and analytical to the prescriptive. The 1992 report of the IPCC contained detailed analysis of the problem and of some of the recommended policies for dealing with it; a further report emerged from the IPCC in 1995. Probably the most useful general text is John Houghton's *Global Warming: The Complete Briefing* (1994, Lion, Oxford).

A useful update of the strategic state of play is provided by Mike Grubb and Dean Anderson's *The Emerging International Regime for Climate Change: Structures and Options After Berlin* (1995, Brookings Institution, Washington DC/Royal International Institute for International Affairs, London). If you want to keep right up to date on the various global negotiations that are underway on emission standards and so on, you could subscribe to the *Earth Negotiations Bulletin*, a version of which is available via the World Wide Web computer network at http://www.iisd.ca/linkages/.

Some of the basic environmental problems facing the world such as climate change are discussed in more technical detail in other books in this series.

3 Sustainable technology

- **Energy resources**
- **End use efficiency**
- **Natural energy flows**
- **Environmental impacts**
- **Energy limits**

All energy technologies have some environmental impacts. To help in the process of selecting which technologies might be most important for a sustainable future, this chapter develops a set of environmental criteria and strategic guidelines. They concern specific local impacts as well as overall energy resource limits. The seven criteria that emerge form the basis of much of the subsequent analysis in this book.

Criteria for sustainability

As we have seen, a range of environmental and resource problems have emerged so far from the use of energy technology. Our review of these issues should have indicated some basic criteria for technologies which avoid or minimise the problems.

Let us assume that our 'technical' goal is *to devise a set of energy technologies which can meet human needs on an indefinite basis without producing irreversible environmental effects.*

From this simple definition of the technical requirement, it is clear that *the technologies should not use fuels which will run out*. This is the first criterion, which is concerned not with environmental impact but with fuel reserves and resource availability. To see what this criterion means in practice, we need to have some idea of what fuel reserves and energy resources are available.

Reserves

Energy resource and fuel reserve issues are clearly of central importance for any sustainable energy system: we have to be able to rely on having continued access to energy sources. It might be reasonable to accept the use of fossil fuels in the interim, while alternative sources are developed, but the fossil fuels do not represent long-term options.

Usable global oil and gas reserves are not known precisely and depend on the price consumers are willing to pay and on the rate of use, but it must surely be only a matter of decades before some key reserves become expensive to access. Coal reserves could last longer, of the order of two hundred years or so.

As we shall see in Chapter 5, **nuclear fusion** is a long way from being a viable option, but it might conceivably provide power for longer periods. So might a switch over to fast breeder reactors which could, in effect, stretch the world's uranium reserves. However, the global availability of the type of uranium used in conventional nuclear reactors is limited to perhaps a hundred years, depending on the rate of use.

Now these are relatively long time scales in everyday terms, although they are very short in terms of human history and even shorter in terms of the geological process of laying down the fossil reserves. In Figure 3.1, energy analyst Gustav Grob puts the fossil fuel era in a wider perspective. He sees it as a short 'blip', to be followed, as fuels run out and/or environmental concerns rise, by a switch to **renewable energy** technology, using natural energy flows. Nuclear technology, whether

Figure 3.1 ***Gustav Grob's view of the long-term patterns in global energy supply***

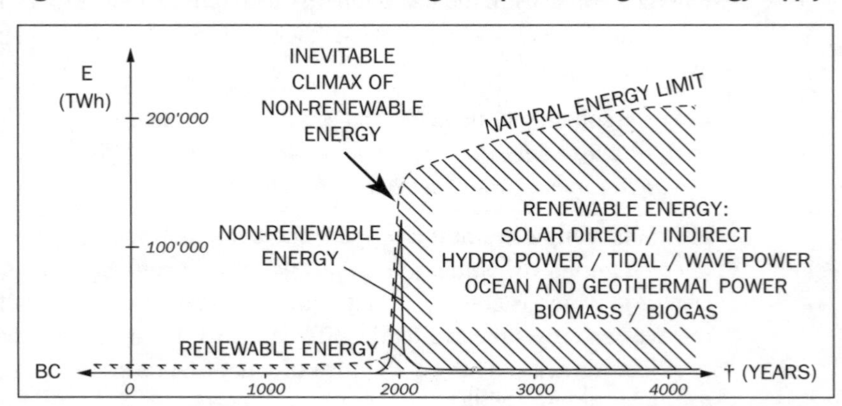

Source: G. Grob, 'Transition to the sustainable energy age', *European Directory of Renewable Energy Suppliers and Sources*, James and James, London, 1994.

fission or fusion, is not seen as relevant on this large timescale; for Grob, renewables are the only significant long-term, non-finite resource (Grob 1994).

However, even on the most optimistic scenario, the change over to renewables would take at least a century, so there would probably be problems with maintaining supplies of some fossil and nuclear fuels (at least, in the absence of the breeder reactor) sometime before renewables could take over fully. As we shall see in Chapter 5, even if it can be successfully developed, it seems unlikely that nuclear fusion could make a significant contribution in time to make up the shortfall. In which case there is a very strong case, just on resource grounds, for using the remaining fossil fuels as efficiently as possible.

So the second criterion is a strategic one: *the efficiency of energy generation and use should be improved as much as possible, as an interim measure while the new sources are fully developed.*

Conserving energy – or generating more?

Some enthusiasts for energy conservation argue that it can achieve so much by way of energy saving at the point of use that the energy supply side becomes more or less irrelevant, at least in the short to medium term. There is talk of savings of up to 90 per cent or more in some sectors.

Certainly if this could be achieved it would be much easier to imagine supplying the small amount left from renewable sources, or even from fossil and/or nuclear sources, while they lasted. The reality however is that it will take time to achieve savings of anything like this level. According to the Intergovernmental Panel on Climate Change (IPCC), at best, if the currently most efficient energy-using devices could be introduced in each ‘end use’ sector, it might be technically feasible to achieve overall energy efficiency gains of 50 to 60 per cent. Although they suggest this might be possible in many parts of the world, they see the timescale as being two to three decades (IPCC 1995).

During that period fossil fuels would still be used. Nuclear power might be able to help reduce carbon dioxide emissions during this period to some extent, but it seems unlikely that it could expand sufficiently to play a very significant role.

The result is that unless there is also investment in renewables, all conservation does is to delay the time when the fossil fuels are burnt off. In the end all the carbon dioxide is still released into the atmosphere, albeit at a lower rate. If the rate of release of carbon dioxide could be dramatically reduced, then of course it might be possible to limit the global warming effect but, as we have seen, by itself energy conservation seems unlikely to be able to achieve this in the necessary time scale.

Using energy efficiently

Although there may be limits to what can be achieved by energy conservation measures on their own, improving the efficiency with which fuel is used is nevertheless vital and urgent as part of a wider strategy for achieving a sustainable energy supply and demand system. This is particularly clear when we look at the way power generation has been carried out so far.

Historically, fuels have been relatively cheap and plentiful so overall conversion efficiency did not seem to matter too much. The giant power stations built during the post-Second World War period in the West were slightly more efficient than their predecessors, but they still only converted around 35 per cent of the energy in the fuel used into useful energy in the form of electricity. Expanding the overall size of the plant can produce marginal increases in conversion efficiency, but at best the maximum efficiency that can be obtained is around 40 per cent. This is not a technical limit but the result of a fundamental thermodynamic constraint associated with the process of using heat to raise steam to drive turbines. It applies equally to steam-raising plant heated by nuclear reactors.

Once generated there are yet further losses of up to 10 per cent or so of the initial power sent out in the distribution process down the national power grid. The efficiency at the point of use is similarly low: for example, many houses are poorly insulated and domestic appliances are often very wasteful in energy terms.

The end result is that only a proportion of the energy in the original fuel is available to be employed at the point of use. The rest is often wasted heat ejected into the environment. The sheer scale of this waste is apparent from Figure 3.2, which shows the energy system at one time projected for the UK, based on a major expansion of nuclear power. As can be seen, around one-half of the input energy would be converted at the power stations into waste heat.

If this trend was continued indefinitely, ever-increasing amounts of heat

Figure 3.2 ***Provision of energy in the UK, projected to 2025 on the basis of the 'official strategy' as conceived in 1976, when a major expansion in the nuclear element was being considered***

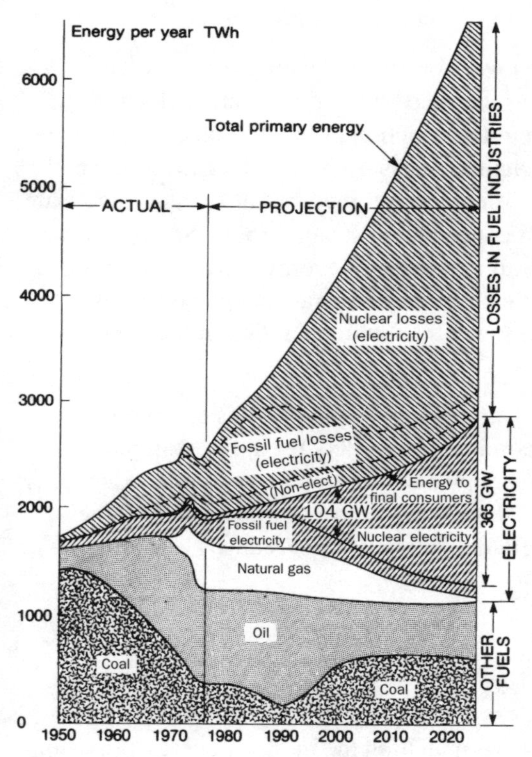

Source: Royal Commission on Environmental Pollution, Sixth Report, *Nuclear Power and the Environment* (The Flowers Report), Cmnd 6618, HMSO, London, September 1976. Crown Copyright, reproduced by permission of the Controller of HMSO.

would be released into the environment. Micro climate effects have already been noted in the immediate locality of some power stations and ultimately, if this process of thermal pollution continued, serious environmental problems could emerge on a global scale, as what has been called the 'heat death' limit was approached. Note that this has nothing to do with global warming: it is just raw atmospheric heating. Nuclear fission or nuclear fusion would be no answer: they release just as much heat.

Combined heat and power

Fortunately, however, use can be made of some of the heat otherwise wasted from power stations – for example, by feeding it to 'district heating' networks, with power stations operating in the so-called 'co-generation' or 'combined heat and power' (CHP) mode. Thus the heat is used as well as the electric power output. Such systems are already in operation around the world, particularly in Scandinavia where many towns have district heating networks, with hot water from power stations being fed to houses and other buildings by underground pipes to provide space heating.

Obviously, given that large power stations are usually built away from centres of human habitation, there may not always be a convenient use for the large amounts of heat available – it cannot be transported great distances. However, CHP does not have to be on a large scale. The most modern forms of CHP involve relatively small and cheap gas turbines, operating something like jet aircraft engines, using natural gas as a fuel. The heat output from such systems is at a higher temperature than that from conventional steam-raising combustion plants, so gas turbines are well suited to CHP and, being smaller, they can be located nearer to heat demands. For example, they can provide heat and power for industrial and commercial use and for housing projects. The overall energy conversion efficiency can be up to perhaps 90 per cent, since most of the energy input to the power stations is converted into useful energy.

Rather than being used in this way for district heating schemes, the exhaust heat from gas turbines can also be used to generate electricity. There has been rapid growth in the adoption of the two-phase 'combined cycle gas turbine' (CCGT). Natural gas is used to fuel a gas turbine and generate electricity, as with normal gas turbines, but the exhaust gases are then used to raise steam for a conventional steam turbine. So both phases generate electricity. The overall efficiency of energy conversion can be up to 55 per cent, much higher than for conventional coal plants or gas turbines plants, although still lower than with operation in the CHP mode.

Clearly, one way or another, gas burning technology has enormous potential and is certainly becoming very widely used: for example, nearly a third of the UK's power plants will soon be generating electricity by burning gas rather than coal. However, most of this involves large CCGT plants rather than CHP: the aim is just to produce electricity (Toke 1995).

Matching supplies to uses

This brings us to the next issue – matching the mode of energy provision to energy requirements in society generally. It seems foolish to use high quality energy in the form of electricity just to heat houses. Certainly this is an expensive option, in part due to the huge losses in conversion described above, which remains true even if combined cycle gas turbines are used. It is still much more efficient to provide domestic heat by burning gas in a modern gas condensing boiler in the home than to use electricity generated in the modern high-efficiency, gas-fuelled CCGT plant. Gas-condensing boilers typically have energy conversion efficiencies of around 85 per cent, while CCGT plants can only achieve around 55 per cent.

This points up our third criterion: *energy production and fuel choices should be matched to the eventual end use*.

In effect this is a logical subset of the second criterion, as is the fourth criterion – concerning actual energy use by consumers. As we shall see when we look in detail at end use energy conservation, devices exist in most end use sectors which require a fraction of the power of currently used equipment. Increasingly energy efficiency criteria are being taken into account in the design of products and production systems – as part of the move to greener products and cleaner production processes. In parallel, buildings are being designed to minimise heat loss and the need for artificial daylighting. Taking all these types of energy saving together, the relevant criterion, our fourth, is clear enough: *energy using technology and systems should be designed to use energy efficiently*.

Energy efficiency makes sense whatever way the power is produced, whether from fossil, nuclear or renewable sources. This is obviously true in straightforward economic terms: it is foolish to waste expensive energy. But it is also true in terms of environmental impact, and particularly in terms of local impacts. Although some power plants may generate less carbon dioxide than others and some generate none at all, there can be local impacts and health risks. The air pollution problems from coal plants are obvious enough, as are the risks of nuclear accidents, but there can also be local problems with the use of renewable sources. The energy may be free, but the land required for the technologies needed to capture diffuse natural energy flows is not. It would be foolish, for example, to cover large areas of land with wind turbines just to power inefficient refrigerators. So the efficient use of power remains vital.

Local impacts of renewables

Given that the use of renewable energy sources might be a key element in a sustainable future, it is worth exploring the issue of potential local impacts in more detail. As you will see in subsequent chapters, this turns out to be a complicated issue and may well determine the success of renewables in becoming a major energy source.

Clearly, technologies that use natural energy flows will not generate pollution or carbon dioxide, but they will have local impacts – such as visual intrusion or the disruption of local ecosystems. At the same time it has to be remembered that if the energy is generated in some other way then the global impacts are likely to be much more significant. Hence the need for some sort of trade-off.

So a fifth criterion emerges, which applies primarily to renewables but also obviously if other energy technologies are being used: *the local impacts associated with energy technologies should be minimised and any remaining local impacts should be traded off against the global environmental benefits of the technology.*

Making such trade-offs is difficult, not least since it is hard to compare the different technologies and types of impact. Comparisons seem to be rather ad hoc and are usually site specific. To try to provide a more coherent framework, a researcher at the Open University, Alexi Clarke, has developed a methodology for examining and comparing the relative impacts of renewable sources. The overall impact depends, pro rata, on the nature and overall scale of the project and the total amount of power produced. However, according to Clarke, in comparative terms, one of the key factors influencing the relative scale of any impact is the proportion of energy extracted from the natural energy flow.

Renewable energy flows

Energy flows in nature in a variety of ways, for example, in the winds, waves and tides, and it is worth considering what happens when we make use of these natural flows. The first point to note is that the extraction of energy from natural energy flows does not significantly affect the overall thermal balance of the planet: the incoming solar energy is simply redistributed to perform other functions. In contrast to the combustion of fuels or the use of nuclear fuels, no extra energy is released into the ecosystem.

Nevertheless, the extraction of energy from natural flows may have a significant impact since some of this energy may have performed crucial functions in the local or regional ecosystem and its release may introduce changes elsewhere in the ecosystem. For example, with hydro schemes located in rivers it is clear that only a proportion of the water flow can be used, otherwise there could be excessive impacts on river life downstream. As can be seen from this example, the key factors influencing impact would seem to be not just the proportion of the natural energy flow that is extracted but also the nature of the local environment. Clarke goes further and argues that the nature of the energy flow concerned is also important, and in particular the energy density (or more strictly the energy flux density) of the flow (Clarke 1994).

This point can be illustrated by looking at a range of technologies and their impacts. Solar collector devices can be used to absorb heat from sunlight falling on them, but they only absorb very small amounts of the diffuse, low energy density, solar energy flow, the result being that the environmental impact is low. By contrast, high head hydroelectric dams attempt to intercept a large proportion of large, high density, energy flows and have correspondingly large impacts on the natural environment. Most of the other natural flow based renewable energy technologies (e.g. wind, wave and tidal power) are strung out somewhere in between these two extremes.

Local environmental limits

The nature of the natural energy flow will need to be taken into account in the process of selecting, designing and locating systems for extracting this energy. In particular, there is a need to relate the way in which specific technologies interact with the natural energy flows in their local contexts (Clarke 1995).

Not all environments will have the same sensitivity to the presence of technologies which extract energy from local energy flows. In some cases the local ecosystem may be able to cope quite well with even high levels of energy abstraction; in other cases it may be that even low levels of abstraction are disruptive.

To complicate matters further, some parts of the ecosystem may be more sensitive than others. For example, in some cases specific animal species are likely to be more sensitive to environmental changes than inanimate geological features, although morphological changes can themselves have an impact on animal and plant life, and in some cases it will be the inanimate part of the environment that is the most sensitive to change.

To explore these issues and assess the significance of any impacts, there is thus evidently a need to distinguish between the various elements in specific local environments – most obviously between the animate and inanimate parts, e.g. animal and plant life and geological structures. These categories can also be subdivided, not least to separate out human beings, who have possibly unique perceptual interpretations of the significance of impacts on the environment, as well as having more direct economic or aesthetic interests in the significance of any changes. You might argue that human interpretations of what represents a 'good environment' are often somewhat biased – for example, toward human concerns such as agriculture, access for leisure pursuits, or for the experience of natural beauty.

In order to progress further with this type of analysis, we would have to begin to consider the conflicts and possible trade-offs between human and environmental interests discussed in the model outlined in Chapter 1. But for our immediate purposes, the analysis can be simplified and we can derive a generalised sixth criterion, as a subsidiary of the fifth: *do not extract more energy from natural flows than the local ecosystem can cope with.*

Overall energy limits

The final criterion for sustainability is the most complex: it concerns overall energy resource limits. Looking back at Grob's chart (Figure 3.1), you will see that he has indicated a natural energy limit for renewables, by which he means the maximum level of energy provision that renewables can supply.

Grob's natural energy limit concept needs some further analysis to see the full implications. Basically it implies an absolute limit to how much energy can be extracted from natural energy flows and processes. Now there are more or less fixed amounts of overall average energy flowing into and around the planet, but the amount that can usefully be extracted will depend on the technology. Currently most extraction techniques have relatively low efficiencies – 10 to 30 per cent or so – but they might improve. That sets an overall technical limit to how much power can be obtained. There are also environmental limits; ecosystems can be damaged if too much energy is extracted from natural flows. Even so, the use of solar, wind, wave, tidal and biomass (i.e. biological sources of energy) will still provide plenty of room for growth – up to and beyond 200,000 TWh p.a. according to Grob, compared to the 100,000 Twh p.a. currently consumed globally. But beyond that there may be a limit to energy availability.

Of course, energy may not be the determining factor limiting human activity on this planet. That also depends on the complex pattern of other interactions with the rest of the ecosystem. So far the ecosystem limits are not fully understood. A whole set of ecosystem related criteria may have to be developed – for example in terms of biodiversity. Even so, although energy availability is not the only factor involved, the natural energy limit concept does at least provide some sort of idea of the ultimate constraints on human activity on this planet and some feel for the role of technology in responding to environmental constraints.

This leads to the seventh criterion, which concerns the nature of human activity on this planet: *technologies should be devised which ensure that human activities stay within the energy limits and carrying capacity of the planet.* This implies that there is a need to decide what are its limits, an issue which will be discussed in more detail later.

Box 2

Criteria for sustainable energy technology

In choosing, developing and deploying new energy technologies there is a need:

1 to avoid the use of fuels which will run out.

2 to improve the efficiency of energy generation and utilisation as much as possible, as an interim measure, while the new sources are developed.

3 to match energy production and fuel choices to the eventual end use.

4 to design energy using technology and systems to use energy efficiently.

5 to minimise the local environmental impacts of energy technologies and to trade off those remaining against global environmental benefits of the technology.

6 to avoid extracting more energy from natural flows than the local ecosystem can cope with.

7 to devise technologies so that human activities stay within the energy limits and carrying capacity of the planet.

Limitations to the criteria

If acted on fully, the seven criteria outlined above and summarised in Box 2 could lead to a shift to a steady state energy system with supply matched to end use demand and impact minimised to be within levels acceptable by the local and global ecosystem. Obviously this is a highly idealised set of criteria: in reality the trade-offs would be difficult and the acceptable threshold levels for impacts are unknown in many cases. Longer term implications are also difficult to predict. Sustainable development is usually defined in terms of ensuring that future generations are not disadvantaged by current activities, but in practice this is very hard to foresee in every case. It is fairly clear that releasing the carbon trapped in fossil fuels is not going to be very helpful, and that bequeathing future generations with nuclear waste to deal with is not going to be well received. But what about tidal barrages, for example? A tidal barrage on, say, the Severn estuary could generate 6 per cent of UK electricity and within a few decades the capital costs would be paid off. So future generations would have the benefit of a cheap renewable source, much like hydro. However, that might not be welcome if it has effects on the local ecosystem.

Tactical and strategic issues like this abound in the sustainable energy field. In Part 2, armed with the criteria, we turn to look at the specific technical options to see how they match up, before looking at some of the tactical and strategic issues in detail in Parts 3 and 4.

Summary points

- There are limits to fossil and nuclear fuel reserves, as well as impacts with using them.
- Consequently, when choosing energy technologies there is a need to consider the efficiency of energy conversion and use and to employ renewable sources wherever possible.
- If renewables are to play a significant role, then the local impacts of renewable energy technologies must be weighed against their global environmental benefits.
- The world's total natural renewable energy resource has limits which may ultimately restrict the level of global economic growth that is possible.
- The criteria developed in this chapter should help in the assessments of the merits of the main sustainable energy options.

Further reading

In addition to the reader from the OU Environment course mentioned at the end of the Chapter 2, the Open University has published a number of specialist's reports produced by the OU Technology Policy Group and analysing the interaction between natural energy flows and human interventions. See, for example, Alexi Clarke's 'Comparing the impacts of renewables: a preliminary analysis', TPG Occasional Paper, 23 December 1993; 'Environmental impacts of renewable energy: a literature review', TPG Report, May 1995.

Some of the more advanced fossil fuel related technologies mentioned in this Chapter (including CHP and CCGT) are described in more detail in another book in this series, *Energy Resources* by Gavin Gillmore.

Part 2 Sustainable technology

Part 2 looks at some of the key technologies that might be candidates for playing a role in an environmentally sustainable energy future. The various energy saving options are examined together with the development of 'green' consumer products and domestic energy conservation techniques and some of the ways in which fossil fuels might be used more efficiently. Other energy supply options are also considered – nuclear power and renewable energy technology.

4 Greening technology

- **Technical fixes**
- **Clean technology**
- **Sustainable energy**

Some environmental problems can be resolved by relatively straightforward technical changes and adjustments – by what are sometimes called **technical fixes**. This chapter looks at some examples of attempts to make 'green products' available to domestic consumers, with energy saving being one of the aims. It also looks at technical fixes in the energy generation field. But as this chapter argues, while some technical fixes may be helpful as a way of reducing energy use and pollution, on its own the technical fix approach may not be sufficient to ensure environmental sustainability. More radical approaches may be needed, especially in relation to energy generation.

Technical fixes

As has been indicated, energy provisions are central to industrial and economic activity – and they are probably the key technologies in terms of environmental impact. The use of energy is determined by activities in sectors other than energy generation, for example, transport, farming, housing, production and so on. Many of the problems with the technologies in these sectors are related to their direct energy use (e.g. emissions from cars), or their indirect use of energy (e.g. pollution from the extraction and processing of raw material used for manufacturing consumer appliances).

Remedial technological solutions exist in most sectors, some of them basically technological modifications or additions designed as a response to the problems created by the existing forms of energy generation and energy using technology. Some of these solutions might be thought of as technical fixes in that they may address the symptoms but often do not deal with the underlying causes. Thus, some technical fixes attempt to clean up the emissions from power plants or cars, but do not change the basic technology which produces them.

In the energy supply sector, remedial technical fix approaches cover each phase of the energy production process, including improvements in the handling of wastes and spoil from mining operations and the efficiency of power plants, and the development of safer oil tankers. Given that fossil fuels will continue to play a major role in the world energy system for some while, even in the most radical alternative energy scenarios, improvements in extraction techniques, plant safety, fuel handling and waste disposal will be of major importance as a way to minimise environmental risks. There will also be a need to develop acceptable ways of storing nuclear wastes. These are key problems facing engineers at present. Some of the relevant engineering techniques are discussed in detail in a companion book in this series, *Energy Resources*.

While finding technical solutions to these problems is obviously important, there is also a need to go beyond ameliorative technical fix measures and to develop more radical technical approaches which might avoid some of these problems. With this in mind and to avoid overlap with *Energy Resources*, rather than exploring the engineering aspects of conventional energy technologies, this book focuses on the development of new, more radical, approaches to energy generation and use. We shall still be looking at a range of technical fixes. Indeed, we will be devoting much of the rest of this book to looking at what they can achieve. However, our emphasis, especially in later chapters, is on more radical longer term approaches – what you might call **sustainable fixes**.

Green consumer products

To start our exploration of technical fixes, let us first look at the some of the developments that have been underway in relation to consumer products. Perhaps the most familiar current examples of technical fixes are the various new environmentally friendly domestic consumer products, many of which are designed to use energy more efficiently: washing machines, refrigerators, TVs, and so on. In parallel new production systems are being developed that are more energy efficient and less polluting. The term 'green product design' is sometimes used to describe the former activities, while the latter is sometimes referred to as being part of a move towards developing 'clean technology', meaning 'clean manufacturing process technology'.

Energy saving is of course only part of the aim: the goal of green product design and clean technology development is wider, involving the reduction of pollution, toxic emissions, and environmental impacts generally. Of course this can often only be partially achieved: rather than using the terms 'green' and 'clean', in reality what we really should be saying is 'greener' and 'cleaner' technology .

As part of this process of developing cleaner and greener technology, there have also been attempts to redesign products to use less materials or to substitute new materials for traditional ones. In part this has been done to reduce the energy used in the manufacture of materials. For example, energy is needed to manufacture the materials used to construct cars, so that cars are said to 'embody' a significant amount of energy beyond what they use directly as a fuel. Generating the energy embodied in a traditionally designed car creates around 10 to 15 per cent of the carbon dioxide emissions that will be produced by burning petrol in the car during its lifetime of operation. To reduce this problem, new less energy intensive materials are used and cars are being designed so that the material can be recycled. Many other consumer products have also been redesigned for ease of recycling, so as to be able to get access to the materials they contain, for subsequent reuse. Here again energy is a key factor: it usually takes much more energy to create new materials from virgin ones than to reuse existing materials.

In general, there is now much more attention being paid to overall 'product life cycles', for example, by making products easier to renovate or repair rather than dispose of and replace, with energy and material conservation being key factors. It is an important part of 'life cycle analysis', the process of assessing the 'cradle-to-grave' environmental impacts associated with the production, use and disposal of products (McKenzie 1991).

Current developments in the green design, materials recycling and clean technology field are the subject of one of the companion texts in this series (Paul Hooper's *Environment and Technology*), so I will not go into more details here, except to consider the implications for energy consumption, some of which may not always be as positive as hoped for.

Energy efficient consumption

As has been indicated, there is a range of technical fixes which focus on reducing energy use by consumers – by offering them new, more efficient, appliances. By purchasing such devices, consumers can have a direct influence on an important element of energy use. Similarly, since it often requires more energy to produce new materials like metal and glass than to reuse existing materials, energy can be saved by recycling materials from domestic wastes and consumers can play a role via bottle and can banks and so on. However, the energy saving can be over-estimated. The energy used in driving to recycling centres has to be taken into account.

Some of the other measures being considered as an aid to energy and material saving in the consumer product sector also need careful assessment. For example, it seems reasonable to press for long-life products rather than products which are thrown away regularly. Yet in some cases consumers might be able to buy new improved products which may be much more energy efficient: given rapid technological advance, in some contexts rapid product replacement may actually be a better way to save energy.

So there may be some technical and operational limits to what consumers can achieve by buying greener products. In some cases, there may even be some contradictions: what might seem to be an environmentally sound step may in the longer term prove to be counter-productive. Of course this is what you would expect of a technical fix, at least in the sense that the term is used in this book. A technical fix implies a technical solution to a social or environmental problem which, while it may solve the specific problem, may create others, or simply push the problem on elsewhere. A more comprehensive solution may involve a fundamental technological change or changes in the way people use technology, in their expectations of it and therefore in consumption patterns. Indeed, it may also require wider changes in economic and social patterns.

However, we also have to be aware of simple-minded 'social fixes' – prescriptions for social changes that will allegedly solve problems easily. For example, it is sometimes suggested that if consumers are given the right information they will buy green products and this will solve at least some of our environmental problems. Consumers can play a role, but it is important to realise that there are overall limits to how much they can achieve via their purchasing decisions by way of, for example, energy saving.

A 1989 study of the purchasing and lifestyle choices made by a typical British family indicated that they could only reduce the amount of carbon dioxide emitted as a result of their activities by around a third. They could choose more efficient domestic heating equipment, buy a more energy efficient car and use public transport more. But most of the rest of the emissions came from activities over which they had no direct control as consumers: for example, the operation of power plants by the companies that supplied them with electricity, and the energy use patterns adopted by industry and commerce. The end result was that even if the family could cut their direct contribution of carbon dioxide production by one-third, the net impact on total energy consumption related to their activities as consumers was only a 10 per cent reduction (Schoon 1989).

This said, consumers can obviously have some influence over the way in which technology is developed. To extend our discussion of technical fixes, let us look at an example drawn from the consumer product area, namely the motor car, since this at least offers the possibility of direct consumer influence via the purchasing decision.

The green car

In recent years there has been growing concern among environmentalists and some car owners about the pollutants that cars emit into the atmosphere. Governments have responded with legislation over air quality and with stricter emission standards. This, in turn, has led manufacturers to develop a number of technical fixes.

One of the first was the development of lead-free petrol – since lead emissions were seen as damaging children's health. The problem is that unleaded petrol is slightly less efficient as a fuel, so typically cars use more fuel and generate more carbon dioxide. Catalytic converters can have a similar effect: they extract some harmful gases but may reduce engine efficiency and certainly they do not reduce carbon dioxide emissions (Nieuwenhurst *et al.* 1992).

An obvious solution is to opt for electric cars. But at present these would have to use batteries recharged with electricity generated by conventional power stations. The emissions of carbon dioxide and pollutants at street level might be reduced, but these emissions would simply emerge at the power station. Given the fact that electricity production in conventional power stations is very inefficient, a system based on electric cars could actually generate more carbon dioxide net than using fuel directly in the car.

The limits of technical fixes can be seen clearly in these examples. So can some possible solutions: for example, the use of renewably generated electricity for electric cars. However, even this is a form of technical fix: the pollution problem may have been resolved, but the result could be queues of electric vehicles. The congestion problem is not addressed, or the damage done by building roads. To resolve these problems would need much more than technical fixes for cars: it would need overall changes in the transport system (e.g. more public transport, more use of trains and buses, more walking and cycling). This might also imply changes in living and working patterns: more use of telecommunications rather than transport; new spatial patterns of residential, commercial and industrial location to make access to work and play less reliant on the car. Air transport may also represent a problem. Air traffic is responsible for around 15 to 20 per cent of the total global warming effect, although, perhaps surprisingly, modern aircraft like the Boeing 777 consume only as much energy per passenger mile on average as modern high-speed trains (Olivier 1996).

Transport policy

The transport issue is a major one and to address it fully would take us well outside the remit of this book since it touches on so many aspects of the environment and public policy. Even if we just stay with the energy related issues, it is clear that the transport question is becoming increasingly significant, and that it will become even more so. Energy use by private vehicles is growing rapidly around the world and is contributing to major air pollution problems, as well as to global warming (Hughes 1993).

As we have indicated, some technical fixes exist, but resolving the transport dilemma is not going to be easy. It has been said that the advent of the car has changed society more than most political changes: so if the car turns out to be a problem then there is likely to be a need for significant social as well as technical changes in order to redeem the situation (Peake 1994).

Given the central role that cars play in the economy of the developed countries, perhaps here is where the conflicts outlined in the model of interests developed in Chapter 1 come into play most visibly. Conflicts between car workers, consumers and shareholders over the relative balance between wages, prices and profits are likely to dissolve in the face of calls for a radical rethink about the use of cars. They are all likely to try to resist change. And yet, everyone has to breathe and consumers are beginning to ask for improvements, as are governments. Even so, any really radical changes in this field will have to be extensively negotiated, which perhaps is why technical fixes are popular, since they seem to avoid the need to face up to major changes.

Energy fixes

The motor car represents a key consumer product and as such, in principle, consumers can influence the nature of the technology by their purchasing decisions. If enough of them want catalytic converters or electric cars, eventually this demand will be met. However, as noted above, this is not true when it comes to the basic technology used for energy generation.

Power plants are large capital items that are designed, funded, built and operated by companies which are relatively remote from direct consumer influence – it is just the electricity that consumers buy. Nevertheless, changes in design do occur, in part due to the imposition by governments of new environmental regulation.

Once again there is a range of technical fixes. For example, one way to reduce emissions from power stations is to switch fuels: burning natural gas instead of coal or oil produces less carbon dioxide per kWh generated. Gas burning has become an increasingly popular option for electricity generation in recent years, not just because of environmental concerns but because gas is relatively cheap and combined cycle gas turbines (CCGT) are very efficient. In the UK, following the privatisation of the electricity industry in 1990, around 20 GW of CCGT capacity has been installed or planned – representing around a third of the UK's electricity generation capacity. Similar levels of expansion are occurring elsewhere.

Unfortunately, switching to natural gas is only a partial solution to the environmental and resource problems. As we have seen, the world's gas reserves will not last forever and burning gas still generates some carbon dioxide. The gas resource issue seems likely to be addressed in the interim in the UK by importing gas from Russia and elsewhere – although reliance on a gas pipeline crossing some of the more politically unstable areas of Eastern and Central Europe raises serious issues in terms of maintaining security of supply. Similar problems will have to be faced elsewhere in the world if gas is to be a major interim energy source for electricity generation.

Energy conservation

Energy conservation at the point of use is a much more effective and longer term technical fix for environmental problems: cutting demand, and therefore reducing pollution. As Table 4.1 illustrates, investment in **end use energy** efficiency is usually much more cost effective than investing in new supply technology, and the potential for energy (and therefore carbon dioxide) savings in most domestic, industrial and commercial end use sectors is considerable.

Table 4.1 ***Energy policy measures in merit order of marginal cost of CO_2 cuts***

In order of cost effectiveness		*Potential CO_2 saving by 2005 (in 10^8 tonnes)*
1	Fuel switching (from electricity to gas for heating)	8.07
2	Electrical appliance improvements	25.97
3	Industrial CHP	20.80
4	Lighting efficiency improvements	32.72
5	Small CHP schemes	6.89
6	Cooker efficiency improvements	4.05
7	Commercial and service sector space heating efficiency improvements	31.63
8	Gas-fired combined cycle power stations	35.28
9	Water heating efficiency improvements	8.63
10	Motive power efficiency improvements	22.92
11	Domestic sector space heating efficiency improvements	34.69
12	Citywide CHP	12.17
13	Process heat efficiency improvements	15.44
14	Renewable energy sources	17.29
15	Nuclear power stations	49.98
16	Industrial sector space heating efficiency improvements	7.77
17	Advanced coal combustion power generation (not CHP)	3.19

Note: Subsequent developments since 1989 will have altered some of the rankings shown here. For example, renewable energy technologies are lumped together, whereas some have now become significantly more cost effective (notably wind power) and should be placed higher on the merit order.

Source: T. Jackson and S. Roberts, *Getting out of the Greenhouse*, Friends of the Earth, London, 1989.

This is not surprising given that the issue of energy waste has only been considered a significant problem in recent years; until recently most energy using appliances have been designed as if energy was almost free. Now, however, technologies for cutting energy waste are emerging across the board. Industry has already made great strides in the development of more energy efficient production systems – not least as a way of cutting costs. But returning to the more familiar consumer product sector, there are low energy refrigerators, cookers, TVs and computers which typically use a fraction of the power of traditional devices, often at only slightly higher purchase cost.

Energy efficient lighting is currently a major area of interest. Low energy fluorescent light can use 70 per cent less power than conventional incandescent bulbs and although such items cost more they last longer and can soon pay back the extra cost in the energy savings they make.

Table 4.2 gives an idea of the relative cost effectiveness of some key domestic energy efficiency options. As can be seen, most of the energy saving techniques are currently all much more cost effective than domestic level energy generation technologies like solar heat collectors. Photovoltaics are even worse: at present they are very expensive.

Table 4.2 ***Simple payback periods (UK) for retrofitting domestic measures***

Renewable or efficiency measure	*Typical domestic savings (£)*	*Simple payback (years)*
Hot water tank jacket	10 to 25	0.5 to 3
Central heating programme/thermostat	15 to 40	2 to 6
Low energy lighting (per compact fluorescent light)	2 to 5	2 to 5
Loft insulation to 150mm	50 to 80	3 to 5
Cavity wall insulation	65 to 140	4 to 8
Draught-proofing	0 to 20	6 to 20
Condensing boiler (full cost gas CH)	20 to 200	6 to 12
Marginal cost when replacing boiler	50 to 80	3 to 5
Solar hot water heating system	20 to 100	12 to 40
External wall insulation	60 to 140	20 to 35
Low emissivity double glazing	40 to 100	60 to 110
Photovoltaic roof tiles	50 to 200	80 to 130*

*Based on Swiss IEA data and multifamily dwelling: single-family homes would be worse.
Source: Ian Bryne, National Energy Foundation, paper to the UK Solar Energy Society Conference 'Renewable energy: what can it do for the environment' 23 Feb 1995.
Data from National Home Energy Rating Scheme/NEF Data; Energy Efficiency Office campaign leaflets; Local Energy Advice Centre database (Bristol Energy Centre); IEA CADDET case studies.

As Table 4.2 shows, many of the best energy conservation measures in cost terms are related to the basic fabric of houses, and energy efficient house design is a key area of development. This is not surprising since residential buildings currently consume around a third of the total energy used in many developed countries.

There are many examples of low energy houses around the world. For example, there are now some dwellings in North America, Scandinavia, Germany and Switzerland which use about 80 per cent less energy than conventional new homes and about 90 per cent less energy than existing homes. In northern climates, the energy consumption for space heating can be reduced by over 95 per cent. This has been achieved by technology which was first demonstrated in Sweden and Canada fifteen years ago, and which came to be called super insulation. Super-insulated buildings have very high insulation levels, energy-efficient windows, draught-proof construction, and a mechanical ventilation system which recovers heat from the exhaust air to preheat the fresh air in winter. There are now several hundred thousand such buildings in Scandinavia and North America, mainly with timber-frame construction. They are also increasingly common on mainland Europe (Olivier 1992).

So it seems that with proper attention to design and insulation houses can run with almost no need for external power supplies other than for cooking and lighting. There are many exotic looking solar houses but Figure 4.1 (page 60) shows a plan for a conventional looking house, which its designers say would nevertheless be a 'zero-fossil fuel, zero CO_2 house'. It is very heavily insulated, and makes use of solar heat via a passive solar conservatory for space heating, and via roof-mounted active solar collectors for hot water. The small requirement for electricity for its low energy lighting and electrical appliances is met, when sunlight is available, by solar photovoltaic cells on the roof: at other times power is taken in from the national grid, but this would be balanced by exporting excess electricity during sunny periods.

Similar developments are occurring in the non-residential buildings sector – in offices and factories, schools and other public buildings, with heat recovery systems, low energy lighting and improved insulation all playing a part (Herring *et al*. 1988).

The limits of conservation

Energy conservation at the point of use clearly has significant implications. However, as with all technical fixes, there are limits; some implementation and operational problems can occur with energy conservation systems.

1 In relation to energy efficient appliances, there is likely to be a delay in deployment, particularly in the domestic sector. Domestic consumers only replace appliances occasionally, when they are old or broken, so it

would take time to exchange the existing range of equipment for more efficient systems.

2 While over their lifetime the more efficient appliances will cost consumers less to run, they may cost more to buy, and this can be a disincentive for those with limited budgets. The old age pensioner in a damp draughty flat may be able to afford a cheap one-bar electric fire but is unlikely to be able to afford a more efficient central heating system.
3 House owners who, due to pressure on jobs and careers, tend to move regularly, may not think it worthwhile to retrofit energy saving measures like insulation in their homes.
4 In addition to these 'social' problems, there can be technical problems. For example, houses designed with very good insulation and air tight double/secondary glazed widows can be stuffy; in some areas there can be a build-up inside of mildly radioactive radon gas, emitted from rocks under the building. In which case some form of powered ventilation may have to be provided – using energy. Ventilation may also have to be provided to avoid condensation.

Designers and technologists can try to minimise and deal with problems such as these and grant aid schemes can be instituted to enable disadvantaged groups to install energy efficient appliances. But it seems clear that while technology can resolve some issues it also sometimes creates further technical or social problems.

Reducing energy demand

Despite the problems discussed above, the potential for technical fixes in the conservation field seems very large, especially if the necessary institutional support structures exist. There is a need for advice agencies to provide information to consumers and possibly for grant schemes, to stimulate uptake. There is also a need for economic assessment frameworks to allow more realistic comparisons to be made between the benefits of investing in energy saving technologies as opposed to investing in new supply technologies.

Increasingly it is being realised that managing energy demand is just as important as managing energy supply, and some new 'demand management' technologies have emerged that may help. For example, some consumer devices, such as freezers, do not need power continuously. They can be disconnected for a few hours during peak energy demand times, thus saving consumer money. Electronic 'smart plugs' have been developed which cut off power from selected devices during peak

Figure 4.1 *The energy showcase: low energy solar house*

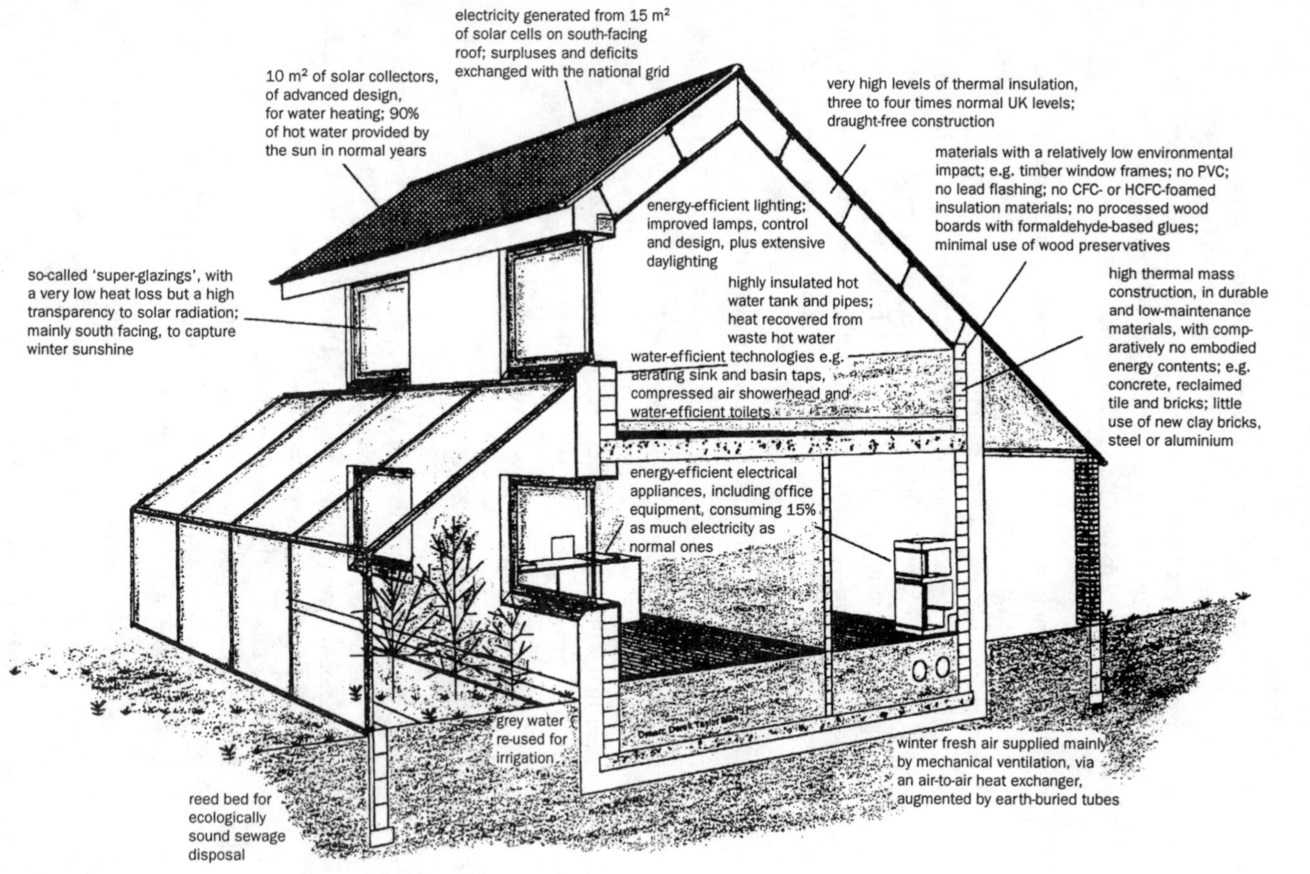

Source: D. Olivier, Energy Advisory Associates, UK.

demand periods when the price of electricity is high. There are countless technical fixes of this sort which could help reduce demand without reducing utility.

As has been indicated earlier, historically the emphasis has been on developing new generation systems: by contrast much less attention has been paid to energy saving and reduction in waste. These priorities clearly must be reversed. It is conceivable that overall energy demand might be reduced by up to 70 per cent or more in many sectors, given time. This would make the problem of finding sustainable supplies of energy much easier and, in terms of our criteria, investment in energy efficiency obviously represents a key element in any realistic sustainable energy strategy (Olivier 1996).

Conclusion: sustainable energy fixes?

Some technical fixes can be useful and important, but others often seem to offer only partial and short-term solutions. For example, switching to cleaner fossil fuels can help reduce emissions, but the only real long-term solution to global warming is to stop burning fossil fuels and to find other more sustainable ways of meeting energy needs. We need to explore these last options in more detail – that is, the more radical longer term options which promise more sustainable solutions. You might call them **sustainable energy fixes**.

Energy conservation is an obvious priority. We have seen that there is a range of technical fixes in the domestic appliance and housing fields and in most other sectors which can help reduce energy demand. But we have also argued that unless non-fossil sources are developed conservation would only slow down the rate of use of fossil fuels. The same total amount of carbon dioxide would eventually be released. A commitment to energy conservation could help slow the rate of emissions, which may help to reduce global warming effects, but there would still be a need to generate power somehow. Conservation, while vital, does not eliminate the need to find new clean and sustainable sources of energy.

The basic issue we must now explore is therefore what are the longer term energy supply options open to us? The options in terms of energy resources are relatively clear. Figure 4.2 illustrates the basic energy incomes and sources that this planet enjoys. Some of them hold out the prospect of supplying power for the long term – the key criterion for sustainabilty developed in Chapter 2.

Figure 4.2 ***Energy: natural sources and flows (in kWh/yr)***

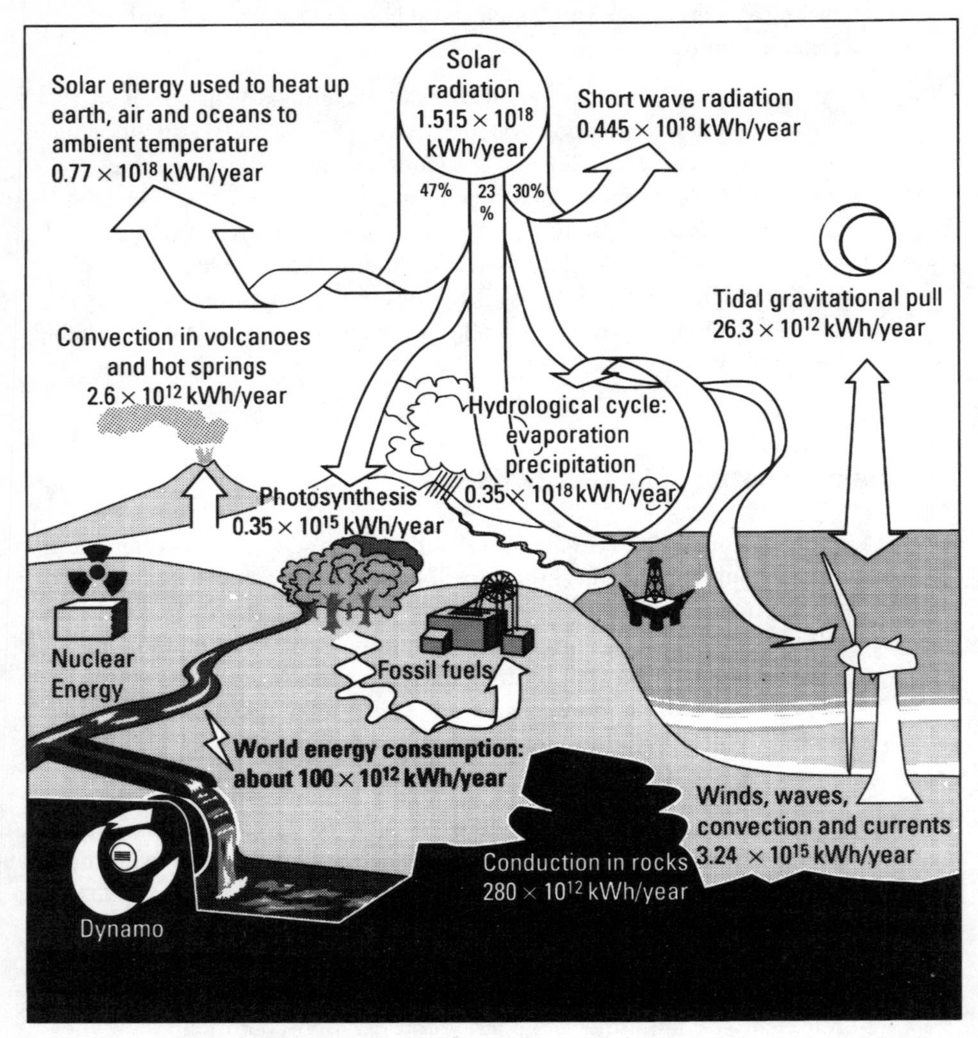

Source: *Physics Review* 2 (5), May 1993: 28. Data from G. Boyle. and P. Harper, *Radical Technology*, Wildwood House, London, 1977.

As can be seen from Figure 4.2, most of the energy sources derive directly or indirectly from the sun. Fossil fuels are really just stored solar energy, trapped underground over the millennia as coal, oil and gas. However, once these reserves have been used up, they are gone for ever. By contrast, most of the so-called renewable sources are based on contemporary solar inputs, providing direct heating or producing winds, waves, water flows in rivers or heads of water in lakes, or short-term stored solar energy in the form of biomass – plant life which can be used as a fuel.

The only non-solar options are tidal power, which is the result primarily of the gravitational force of the moon acting on the seas, geothermal energy deep in the earth, and nuclear energy, from the break-up of the nuclei of atoms. Radioactive decay from natural nuclear processes within the earth is what supplies some of the geothermal heat, but nuclear heat has also been released artificially, by **nuclear fission** and **nuclear fusion**. Nuclear sources are not indefinitely sustainable, but their proponents argue that some of them can nevertheless supply power long into the future.

To summarise, if more radical and sustainable technical fixes in relation to new energy supplies are required, the only non-fossil options known of at present are nuclear fission, nuclear fusion and the various forms of renewable energy. The next chapters look at these options in detail, starting first with nuclear power.

Summary points

- Technical fixes and product redesign can help reduce environmental impacts, for example, by improving energy efficiency.
- In the energy sector, energy conservation and the more efficient use of fossil fuels are obvious priorities.
- However, if full sustainability is to be achieved then there may be a need for more radical technical fixes.
- There is a range of potentially sustainable energy fixes including renewable energy technologies and nuclear power.

Further reading

The development of green products and clean technology is one of the central concerns of another book in this series, Paul Hooper's *Environment and Technology*.

A good starting-point to explore the idea of clean technology in more detail is R. Kirwood and A. Longley (eds), *Clean Technology and the Environment* (1995, Chapman and Hall, London). The January–March 1996 issue of *Co-design* (5–6) is devoted to green design and contains some good introductory papers.

The Green Car Guide, P. Nieuwenhurst, P. Cop and J. Armstrong (1992, Green Print, London), contains some very useful material on the problems associated with the use of cars and some of the technical fixes that have emerged. Daniel

Sperling's *Future Drive* (1994, Island Press/Earthscan, London) looks at some of the electric car technical fix options in more detail, as does James MacKenzie's *The Keys to the Car* (1994, World Resource Institute/Earthscan, London).

Technical fixes in energy generation and use are discussed in Dave Toke's *The Low Cost Planet* (1995, Pluto, London) and in most contemporary energy texts. Amory Lovins of the Rocky Mountain Institute in Colorado (see Appendix II) is an energetic exponent of improved energy efficiency: see his many publications, starting with his seminal book *Soft Energy Paths* (1977, Penguin, London).

For a good general discussion of the limits of technical fixes see Barry Commoner's *Making Peace With the Planet* (1990, Pantheon Books, New York).

5 The nuclear alternative

- **Nuclear fission**
- **Nuclear waste**
- **Fast breeder reactors**
- **Nuclear fusion**
- **A nuclear future?**

Nuclear power provides nearly 6 per cent of the world's primary energy at present. Given that its use does not generate carbon dioxide, this chapter asks whether nuclear power can be relied on to play an increasing role in combating global warming, or whether, as many environmentalists would prefer, it should be phased out. We also take a look at nuclear fusion, which some people see as a possible energy option for the longer term.

Nuclear power

The basis of the human use of nuclear energy is the fact that the nuclei of the atoms of some naturally radioactive materials found in the earth, most notably one of the isotopes of uranium, can be induced to split up or undergo **nuclear fission**. In doing so large amounts of heat are produced, along with nuclear radiation and a range of radioactive fragments from the fission process. The heat can be used to boil water to raise steam, as in a conventional fossil fuel fired power plant. If the fission or splitting up process can be made to happen very rapidly then an explosion occurs – the atomic bomb. In conventional nuclear reactors the physics of the design is such that this cannot happen, but in theory this is an outside chance in some of the more advanced reactors like the fast breeder.

There is a range of technologies for turning nuclear heat into electricity. In the USA the emphasis has been on water cooled reactors – in part because the first power reactors were designed for use in submarines and had to be compact. Gas-cooled reactors, as initially developed in the UK, tend to be physically larger and, it now seems, more expensive. Most of the world's 430 or so reactors are of the water-cooled type, chiefly based on the pressurised water reactor (PWR) which was initially developed in the USA. Typical modern nuclear reactors range up to around 1,300 megawatts of generating capacity.

Nuclear technology is complex and what follows must necessarily be a simplified account. However the basic concepts are fairly easy to describe. Figure 5.1 shows, in simplified form, the internal layout of a typical PWR, illustrating how the heat energy is absorbed from the reactor core by a pressurised water cooling loop. This in turn transfers heat via a heat exchanger to generate steam for power production, as with a conventional fossil fuelled plant.

All nuclear reactors produce nuclear wastes, many of which are fiercely radioactive. They are the material results of the basic fission process. The most radioactive products of the fission process tend to 'decay', that is lose their power, relatively quickly. But some of the less radioactive materials can remain active for very long periods. For example it takes

Figure 5.1 ***Simplified plan of a pressurised water reactor***

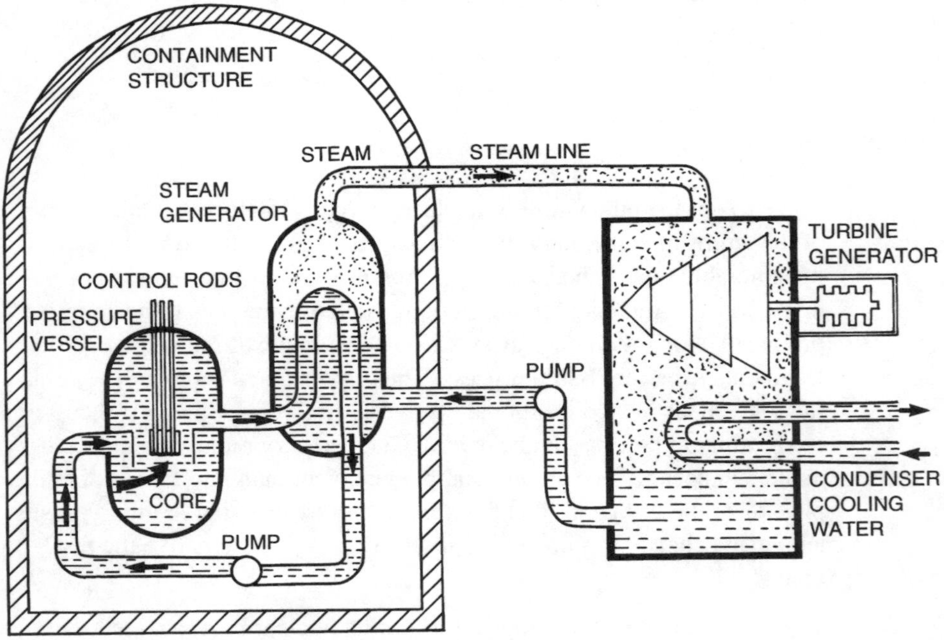

Source: Sally Boyle, Open University, 1996.

around 25,000 years for the activity in plutonium to fall by a half, the so-called 'half life'. It would take even longer for the activity level to fall off to negligible levels – say ten half lives, or around 250,000 years or more.

Plutonium is the key ingredient of nuclear weapons, and it can also be used to fuel fast breeder reactors, which convert otherwise 'non-fissile' (i.e. non-fissionable) uranium into plutonium, thus generating or 'breeding' some new fissile fuel. To get access to the plutonium produced by fission processes, the old ('spent') fuel from conventional reactors, or the fuel from breeder reactors, must be chemically treated to separate the plutonium out from the other nuclear wastes. This process is messy and generates secondary wastes. Indeed this reprocessing activity is what generates the bulk of nuclear waste, even if it is at lower levels of activity than the original wastes.

Radioactive materials and the radiation from them can be very dangerous for living things (see Box 3), and extreme care is taken to ensure that they are not released. Government controls are very strict and the engineers and technologists involved have tried their best to make the systems foolproof. But no technology can be 100 per cent safe, and accidents can always happen. The chance of major accidents is very low, and can be reduced if sufficient money is spent on safety. But there still remains a chance of incidents, and the impacts can be very severe.

Nuclear accidents

A review of the potential impacts of energy technologies, including nuclear plants, published in the journal *Nuclear Energy*, noted that 'for a 10 per cent core release from a typical PWR, total mortality of approximately 10,000 people is expected in a typical Western European location with appropriate counter measures'. It added that 'the probability of such accidents is obviously open to speculation. Some industry sources put the probability as low a 1 per 10^8 (100,000,000) GW years for a PWR. Actual world operating experience to date with a wider range of reactor types is closer to 1 in 10^3 (1,000) GW years' (Eyre 1993: 324).

These odds sound reasonably long. However, with, say, 100 reactors rated at 1GW each, the chance of a significant accident increases significantly – to once in every ten years, on the basis of the 10^3 GW years figure. Statistical chances are unreliable guides to reality, and, so far, there have been only three major nuclear accidents – a fire at the Windscale plutonium production plant in the UK in 1957, the Three Mile

Box 3

Impacts of radiation

Some forms of radiation occur naturally (from rocks and from cosmic rays coming in from space) and the nuclear industry argues that, on average, its activities add only small extra amounts to what most people already experience. However, even though great care is taken to avoid exposing people to them, some of the products of manmade nuclear fission are unique, not least in their intensity.

In addition to producing heat, the nuclear fission process splits fissile materials like uranium or plutonium into radioactive fission products and also leads to the emission of very energetic nuclear particles (neutrons and protons) and electromagnetic radiation (gamma rays – a more powerful form of X-rays). All these products of the fission process can have serious impacts on living material. High energy nuclear particles or gamma rays can damage the material in cells, particularly young, rapidly growing cells. This is why radiation treatment at low levels is sometimes used to treat some cancerous growths. However, uncontrolled exposure at a higher level can cause cancer, and at very high levels it can kill outright.

Some of the radioactive materials produced by the fission process are not very radioactive, but they have long half lives, and this combination can be a problem. For example, the radiation from plutonium is relatively weak and can be absorbed by a few layers of human tissue. That means if a person accidentally ingests or breathes in a small speck of plutonium and retains it in the body, its presence cannot easily be detected from outside the body, since the radiation is shielded by the body material. However, cells near to the site of the plutonium are continually irradiated, and eventually cancers may be produced.

Clearly, exposure to all forms of radiation can be dangerous. The main forms of protection are physical barriers, to absorb the radiation; physical separation by distance; and limiting exposure time. Thus nuclear reactors are surrounded by shields to absorb radiation; the length of time any operating staff are exposed to radiation and the degree of its intensity are carefully monitored; reactors are usually sited in remote places to minimise risks to local people in the event of an accidental release of radioactive material.

Island accident in the USA in 1979, and the Chernobyl accident in 1986. Given that there are now around 430 reactors operating around the world, we would seem, so far, to have been statistically fortunate.

The fire at the Windscale plutonium production pile in Cumbria resulted in the release of radioactive material, including around 20,000 curies of radioactive iodine (one curie equals the radioactivity of one gram of radium). Crops over a 300 sq. mile area were destroyed as a precaution, and two million litres of milk were poured away (Arnold 1992). The near melt-down accident at Three Mile Island was contained, although there were some releases of radioactive material and at one time mass evacuation of at least 630,000 local people seemed imminent (Garrison 1980). In neither case were there any direct casualties, although there have been suggestions that some early deaths might have occurred subsequently in the population downwind from the Windscale plant (McSorley 1990). The Chernobyl accident, by contrast, was much more severe, and led to concern right across the world as the radioactive plume spread across North-West Europe and even, although much weakened, to North America. The accident occurred during an attempt to check the plant's safety system, when it was running at very low power, and only a proportion of the radioactive material it contained was released into the atmosphere (Read 1993). Even so, quite apart from thirty-one immediate deaths on site and among firefighters, there have been reports of subsequent deaths and serious illness in the region around the plant. Estimates vary from a few hundred early deaths to several tens of thousands, while some reports suggest that, ultimately, the health of several million people may be adversely effected (Scorley 1990; Medvedev 1990).

Obviously, these figure have to put in context. Major dam failures can also kill many people and supporters of nuclear power often claim that, compared with its alternatives, nuclear power is a relatively safe option. This is not the place to enter into the complexities of statistical 'risk analysis' (for an interesting introduction see Perrow 1984). However, it does seem that nuclear accidents are widely perceived as different in kind, not least because they can lead to illness many years later, and concern over the potential long-term consequences of accidents has been one reason why nuclear power has been opposed by most environmentalists.

A failed option?

Despite the risks, nuclear power might be seen as the ultimate technical fix. Large amounts of energy can be released from small amounts of relatively abundant material. Certainly in the period after the Second World War it looked very promising: there were even suggestions that nuclear electricity would be 'too cheap to meter'. The reality has proved somewhat different. Fifty years on after very large investment around the world, it still only provides under 6 per cent of the world's primary energy and the cost of the electricity produced remains high. For example, in the UK, following the privatisation of the electricity industry in 1990, electricity consumers had, in effect, to pay a surcharge of around 10 per cent to meet the extra cost of the 20 per cent or so of the UK's electricity which nuclear plants were supplying.

Economic problems coupled with public opposition have meant that, in many countries nuclear power programmes, private or state run, have been winding down or have been halted. Nuclear plants provide around 20 per cent of the USA electricity but the accident at Three Mile Island effectively halted expansion. Denmark decided not to go nuclear, following a national debate during the late 1970s followed by a referendum, and the Chernobyl disaster led several European countries with existing nuclear programmes to halt further expansion or to develop phase-out plans. For example, Italy decided to abandon nuclear power in 1988, and Sweden is trying to phase out its nuclear plants. While attempts are being made to keep the nuclear programme going in Eastern and Central Europe, and there are plans for expansion in some countries in the Far East, within Western Europe only France has retained a major commitment to nuclear expansion.

In the UK a moratorium was imposed on nuclear developments following the first failed attempt to privatise the industry in 1989–90. The arrangements for privatisation outlined in the Government's 1995 White Paper mean that any further expansion in the UK will have to be privately funded and this seems unlikely. In November 1995, British Energy, the new private company which is to take over the bulk of the UK's nuclear plants, indicated that it had no plans to build new plants.
As a consequence the industry is faced with slow decline, from the high point of the 30 per cent or so contribution to UK electricity achieved following the completion of the Sizewell BPWR. As the existing older plants reach the end of their working life and are closed down the contribution will drop (see Figure 5.2).

However, nuclear reactors do provide a way of generating electricity without producing CO_2 (or SO_2) and so around the world there are still some signs of interest in the nuclear option as one possible response to **global warming**.

Nuclear power and global warming

Nuclear plants generate electricity and produce no conventional pollution or greenhouse gases like carbon dioxide. On this basis the industry argues that it should be supported as a way to help limit global warming and **acid rain**. How valid is their argument?

The UK Atomic Energy Authority has suggested that, given massive backing, it might be possible to increase the global nuclear contribution by nearly threefold, up to around 50 per cent of world electricity

Figure 5.2 ***Projected output from Nuclear Electric's plants in the UK***

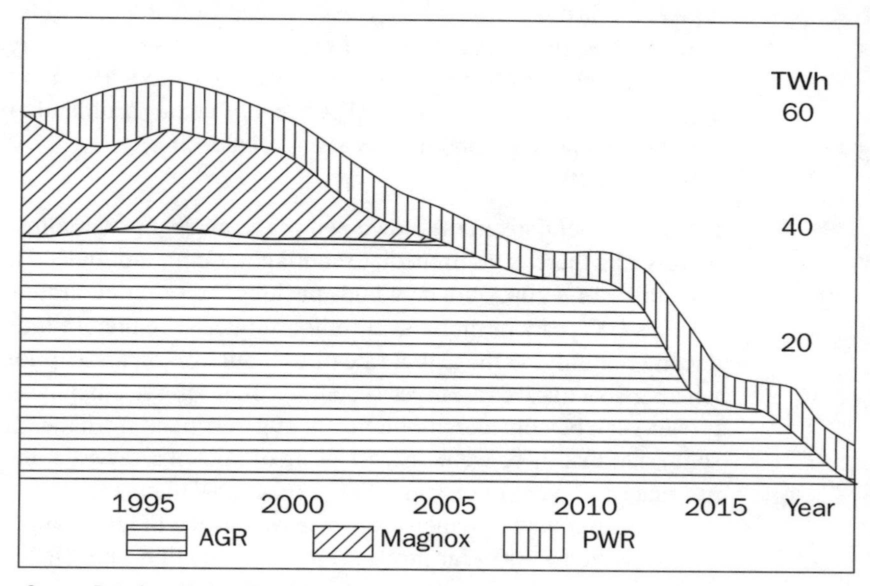

Source: Data from Nuclear Electric's evidence to the UK Government's Nuclear Review, 1994.

requirements by the year 2020, which would result in a 30 per cent reduction in carbon dioxide emissions (Donaldson and Betteridge 1990: 26). However, not only would this be very expensive, but there are also other major drawbacks. The most obvious is that all nuclear plants generate dangerous nuclear waste, some parts of which remain dangerous for thousands of years. Some countries have developed repositories for high level waste in remote areas, and there are vitrification techniques for converting some wastes into a glassified form. But no one can be sure if the wastes can be successfully contained over such long lengths of time. Few communities are willing to accept waste repositories near them, and yet more and more wastes are being produced. This is a real problem which many environmentalists feel can only really be tackled if no more waste is created.

Uranium reserves

The next point is that the nuclear option may also be relatively short term – uranium reserves are not infinite. In 1990 the UK Atomic Energy Authority (UKAEA) suggested that if an attempt was made to expand nuclear power dramatically on a worldwide basis in response to the threat of global warming using conventional 'burner' reactors, 'the world's uranium supplies that are recoverable at a reasonable cost would be unlikely to last more than fifty years' (Donaldson and Betteridge 1990: 29). The UKAEA's journal *ATOM* was even more specific, carrying an article in 1990 which suggested that 'for a nuclear contribution that expands continuously to about 50 per cent of demand, uranium resources are only adequate for about 45 years' (Donaldson *et al.* 1990: 19).

For the moment, given the slowdown of the nuclear programme worldwide, there is no immediate shortage of uranium: indeed there is something of a glut, with new finds pushing the resource limit up to perhaps 100 years or more, at current use rates. The precise figure will depend not only on the actual rate of use, but also on assumptions about whether speculative resources (i.e.anticipated reserves that have yet to be proven) can be considered reliable. A 1991 estimate from the UKAEA suggested that there were only 81 years of known reserves at then current use rates, although there were 333 years of speculative reserves (Eyre 1991a). But finds since then seem to have pushed the figure up: a 1993 review quoted a 100-year proven reserve/production ratio (Jones 1993).

However, as the quotes from the UKAEA make clear, shortages would occur if a major nuclear expansion programme was launched. Indeed, Brian Eyre, then Deputy Chair of the UKAEA, noted that if the proportion of the world's electricity that was supplied by nuclear power was increased from 16 per cent as at present to 20 per cent, just using conventional burner reactors, then 'all the uranium resources, including the speculative resources, are committed by 2060, but if the nuclear fraction increases to 50 per cent, then the uranium is committed by around 2025' (Eyre 1991b: 12).

It might be that, as reserves of high quality uranium ore became scarce, they could be stretched by using lower grade uranium ores. However, the net energy balance might not be favourable: uranium ore contains only a very small amount of the type of radioactive uranium needed for fission reactors (the uranium 235 isotope) and this has to be extracted and enriched to make usable fuel through a very energy intensive process. Using lower grade ores would make it even more energy intense. At some stage the so-called 'point of futility' is reached when, as lower and lower grades of uranium have to be used, more power is required in order to process the fuel than can be generated from it.

One study suggested that if a major nuclear programme was launched in response to concerns about global warming, then at least initially, more carbon dioxide was produced net than if the electricity from fossil fuel plants was simply used by consumers directly. The reason is that, although nuclear reactors might eventually provide the energy for nuclear fuel processing, in the initial phase the energy for processing the increasingly lower grade uranium ores would have to come from fossil fuelled power plants (Mortimer 1990).

There are some other options. If uranium became scarce, some of the world's existing plutonium supplies, including the plutonium from redundant nuclear weapons, could perhaps be used as a fuel – mixed with uranium – in conventional reactors. Some people have welcomed this idea as a way of disposing of plutonium and using bombs for better purposes. However, the process would generate further nuclear wastes, including more plutonium, and at present, given the glut of uranium, there seems little commercial interest in the idea.

Fast breeder reactors

More dramatically, uranium reserves could be used more efficiently in **fast breeder reactors** (FBRs). Fast breeders are reactors which can 'breed' plutonium from the otherwise wasted parts of the uranium. In principle they can extract fifty to sixty times more energy from the uranium, so that the availability of fissile material could, in theory, be extended by up to fifty to sixty times. Using figures like this, and assuming current reserves are put at 100 years, enthusiasts sometimes talk in terms of uranium reserves being stretched thereby to 'thousands' of years, although more guarded commentators limit it to around 1,000 years (Eyre 1991a).

Certainly, the claim that uranium reserves could be extended in practice by a factor of 50 could be seen to be somewhat optimistic, especially since the discussion concerns as yet hypothetical systems involving networks of fast breeder reactors and reprocessing plants. If this proved to be the case, the result could be that, in reality and assuming a dramatic increase in nuclear power generation in response to global warming, uranium reserves might still only last for a few hundred years, even using fast breeders, so 1,000 years might be an optimistic estimate.

In this context, it is important to realise that it would also take time for a breeder programme to make a significant contribution in energy terms. Despite the name, the breeding process is not 'fast'; it can take years to breed the same amount of plutonium that was started with. The so-called 'doubling time' can be in excess of twenty years, perhaps up to thirty years, especially when the fuel cooling, reprocessing and refabrication processes are taken in to account (Mortimer 1990). The word 'fast' simply refers to the type of nuclear interaction that occurs in breeder reactors: it involves fast, high energy neutrons, rather than the slow, low energy neutron interaction in conventional reactors.

In a breeder, the fast neutrons convert the non-radioactive part of uranium into plutonium, a process which also occurs, but much less efficiently, in conventional reactors. As was noted above, to get access to this plutonium the old 'spent' fuel must be reprocessed, so that a fast breeder based system would require significantly enlarged reprocessing facilities. Not only would this increase the amount of nuclear waste produced, but it would also, presumably, increase numbers of shipments of spent fuel and plutonium which would have to move between the various reactors and reprocessing plants. This would open up a whole range of safety and security problems, not least the risk that, despite all precautions, plutonium could be stolen for bomb-making activities.

So, although the fast breeder option has some attractions and could extend the lifetime of the uranium resource, there is a range of associated problems. Despite enthusiasm from the nuclear advocates, who realise that the breeder is the only realistic future for fission, in practice the prospects for breeders are unclear. Around the world fast breeder projects have been shut down, for example, in the UK, Germany and the USA. In the US, President Carter was evidently concerned about the potential security problems of plutonium proliferation and a moratorium was imposed on the fast breeder programme in 1977. In Germany the prototype fast breeder at Kalkar, the scene of major protests by objectors in the late 1970s, was finally abandoned in 1991. The UK government was alarmed by the costs of the FBR project at Dounreay and at the long time scale before a commercially viable technology might emerge; the programme was halted in 1994. Japan and France retain significant FBR programmes but these have had technical problems, most recently (in December 1995) with an accident at Japan's Monju plant, which led to a review of the country's nuclear programme.

Nuclear fusion

The only other significant nuclear option is **nuclear fusion**. Rather than causing atoms of uranium or plutonium to split, as in nuclear fission, in a fusion reaction, atoms of a form of hydrogen can be induced to fuse together at very high temperatures (tens of millions of degrees centigrade) to form helium. This process is accompanied by the release of vast amounts of energy, much more than from fission. This is how the sun works. It is also the principle of the hydrogen bomb, where the initial high temperature is created by exploding an atomic bomb, which then triggers a fusion reaction.

Theoretically, the process can also be slowed down and carried out under controlled conditions, for example, by holding a plasma of very hot gas in a vacuum away from containing walls by the use of magnets. To achieve fusion in these conditions very high temperatures are required – around 200m °C. If the fusion reaction can be sustained the fusion reactor could, in theory, provide heat for power generation. Progress is being made with experimental devices like JET, the Joint European Torus at Culham in the UK but, so far, no fusion reaction has been sustained for more than a very short time. A more advanced machine, the so-called International Thermonuclear Experimental Reactor, is planned at a cost of £7.3 billion, in an attempt to get nearer to the conditions for a sustained fusion reaction. However, a full-scale power producing reactor is generally seen as, at best, decades away. If such a reactor could be built then, in principle, fusion would be better than fission in resource terms since, instead of uranium, it uses materials which are relatively abundant. The basic fuels in the most likely configuration to be adopted would be deuterium, an isotope of hydrogen which is found in water, and tritium, another isotope of hydrogen which can be manufactured from lithium. Water is plentiful but lithium reserves are not extensive: even so it is claimed that they might provide sufficient tritium for perhaps 1,000 years, depending on the rate of use (Keen and Maple 1994).

Fusion reactions of the type likely to be used in reactors create large amounts of high energy neutrons which would have to be trapped, thus collecting energy from the reaction: that is how power would be generated. Although fusion does not produce direct wastes other than helium gas, the containment materials and equipment in a fusion reactor would become fiercely radioactive and have to be stripped out and stored periodically. These activated materials would only have half lives of around ten years, so that their radioactivity would decay more rapidly than that in some wastes generated in fission reactors, but these materials would still remain dangerous for perhaps 100 years (Keen and Maple 1994). There would still therefore, in effect, be a waste problem.

There could also be safety problems regarding the fusion reactor. Fusion reactions are difficult to sustain, so in any disturbance to normal operation the reactor would shut down very rapidly. But it is conceivable that some of the radioactive materials might escape if, for example, the superhot high energy 'plasma' beam accidentally came into contact with and punctured the reactor containment before the fusion reaction died off. The main concern is the radioactive tritium that would be in the core of the reactor: the volumes would not be great but any escapes could be very serious.

A practical reactor?

Finally, there is still a need to find a way of extracting useful energy out of a fusion reactor. At first sight this looks like an impossible task. On the one hand there is a plasma at 200m °C and on the other, assuming a convention power engineering approach, water needs to be boiled to generate steam for a turbine. The mismatch seems vast: a hot water pipe cannot be put through the plasma. Fortunately, however, it is not the heat of the plasma that would be tapped. Rather it is the energy in the intense neutron emissions that emerge from the fusion reaction that would be absorbed in some way and converted into heat, this in turn presumably being extracted by a conventional heat exchanger.

Given that it will be some time before a workable fusion reactor is available, not too much effort has been put into developing ideas for exactly how it might be used for power production. At some point it may be possible to convert the energy emerging from the fusion reaction directly into electricity, but for the foreseeable future if a fusion reactor can be built it will, perhaps rather strangely given its high tech nature, still have to rely on a traditional steam-raising boiler to generate power.

Despite very large scale funding over the years (some £20 billion has been spent worldwide so far) there is some way to go before fusion can be seen as anything more than a long shot option. The physics has still to be fully resolved and a workable commercial device is at best decades away. It might therefore be considered to be irrelevant to current energy and environmental concerns. As has been indicated, there is a range of operational problems and as yet few people would hazard a guess as to the economics of such systems. On this basis, there have been objections to the level of funding that has been given and continues to be given to fusion research.

In terms of the strategic criteria outlined in Chapter 3, the fuel resource for fusion seems reasonably long term. However, by the time a workable system has been developed, if it ever is, the world's energy problems will need to have been solved already. In which case, while it might be reasonable to continue some research as a long-term insurance, it could be argued that given the inevitable scarcity of funds and the urgency of our environmental problems, rather than trying to build a fusion reactor on earth, it might be more effective to make use of the one mankind already has – the sun.

Dis-invention

In terms of the criteria outlined in Chapter 3, of the nuclear options only the fast breeder and fusion represent technologies which might in principle supply power for significant lengths of time. Neither fusion or the breeder can offer indefinite supplies, but in both cases the energy resources, although not renewable, seem relatively large, of the order of 1,000 years. However, this depends on the assumptions made about the productivity of the technology and the rate of fuel use. While enthusiasts sometimes claim higher estimates, 1,000 years might turn out to be optimistic, especially if the nuclear contribution was expanded dramatically. Moreover, as has been noted, there is a range of operational and strategic problems which could limit the acceptance of these technologies and certainly the scale of any contribution in the immediate future seems likely to be small. In practice neither seem to be very relevant to the actual situation at present in terms, for example, of responding significantly to global warming.

Although the nuclear option may not seem to be very relevant in the immediate future, it still absorbs large amounts of research funding. For example, the European Commission allocated Ecu458 million to its four-year 1990–94 fusion research programme, as against only Ecu157 million to its non-nuclear energy programme.

Over the years this level of funding has helped the nuclear lobby to build up a powerful institutional base. However, priorities are changing. For example, research budgets, at least for fission, have been falling (see Figure 5.3) and within the nuclear lobby there are signs of awareness of a need for a new approach, in effect recognising that some past expectations may have to be abandoned or revised.

In a paper entitled 'What now?', presented to a conference on the 'Next Generation of Nuclear Technology' at the Massachusetts Institute of Technology in October 1993, Professors M. Cohn and L. Lidsky, both members of MIT's Department of Nuclear Engineering, concluded that:

> For the time being at least, nuclear power has exhausted its privileged research funding position. Opportunity costs considerations require that competing technologies receive serious funding. Only if these options fail can a case be made for renewed flows of disparately large amounts of public funds to nuclear power.

They suggested that there could be an interregnum, that is a period of low activity, in terms of nuclear R&D, but felt that:

> In some respects the interregnum, if there is to be a second nuclear era, may turn out to be a blessing in disguise for nuclear power. Many scholars have written about the problems of bureaucratic inertia in general and technological lock-in with respect to Light Water Reactors in particular. Even as the nuclear industry winds down, sunk costs in the commercial sector and interlocking subcultures in government bureaucracies and academia, inhibit innovation. A clean break offers the opportunity of a fresh start unburdened by intellectual and emotional bias.
>
> (Cohn and Lidsky 1993: 7)

The paper argues against the more conventional view that it is vital to keep expertise alive, by suggesting that 'it may well be that we do not want to preserve our current knowledge' in that 'it is not clear that

Figure 5.3 ***UK nuclear research and development allocations 1986–95***

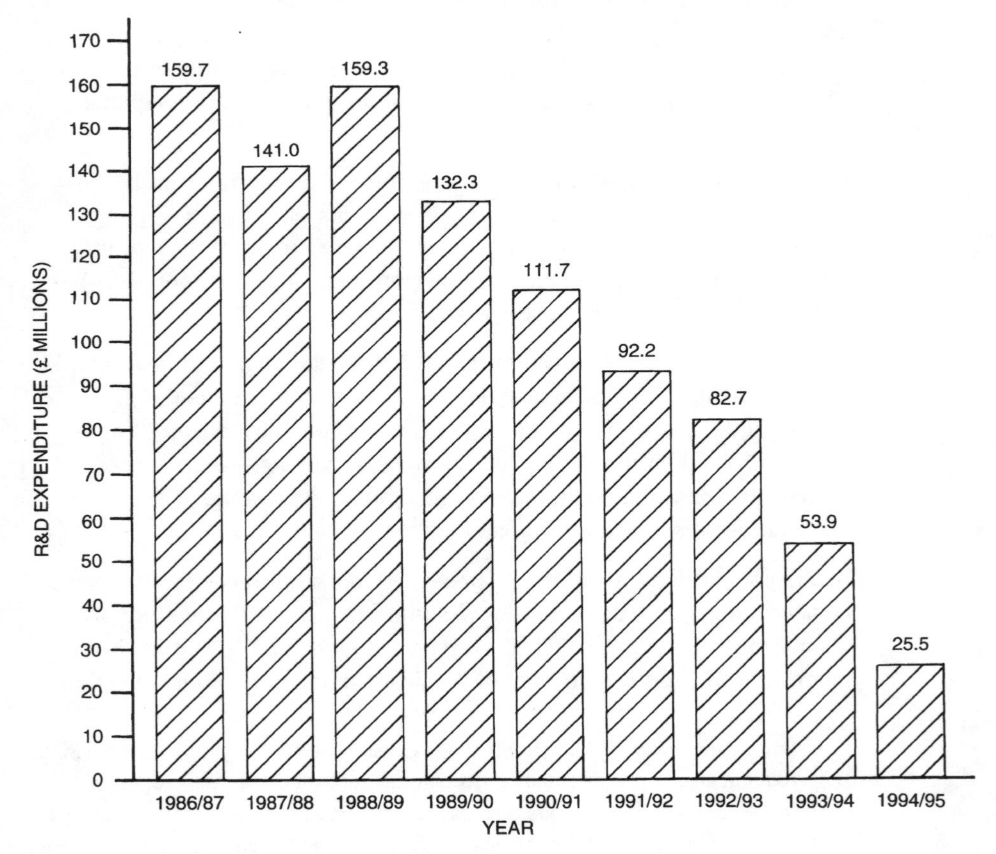

Source: *The Department of Energy's Nuclear R&D Programmes: A Consultation Document*, Department of Energy, August 1991. Crown Copyright, reproduced by permission of the Controller of HMSO. Updated (for 1990–95), with data (including estimated outturn for 1993–4 and provisional allocation for 1994–5) from the Office of Science and Technology's 'Forward Look', Statistical Supplement Table 2.18.2(1), Cabinet Office, HMSO, 1994.

current nuclear expertise will be relevant to nuclear projects 30–50 years from now' (Cohn and Lidsky 1993: 4). As can be seen, the authors are in effect prescribing a process of what amounts to 'dis-invention', or perhaps dis-innovation, with key elements in the institutional support structure for the existing nuclear research trajectory being dismantled in order to allow alternatives, whether nuclear or otherwise, to develop.

Even relinquishing nuclear research completely would not necessarily mean abandoning nuclear power. The nuclear option would still be there as an insurance if it proved to be needed in the future, for example, if the renewable energy technologies could not be developed fully for some reason.

Moreover, for the present at least, there is still the existing network of nuclear power plants. So even on the most radical view, nuclear power will be with us for some time. This said, although the nuclear industry obviously still has aspirations for expansion, equally there are strong pressures for a shift away from nuclear power and, more radically, for a rapid nuclear phase-out.

A 'dry run', as it were, is already underway in the UK and elsewhere, at least in terms of gaining expertise in the technical problems of decommissioning nuclear plants – as older plants reach the end of their working life. Decommissioning is expensive and can generate extra waste to be stored, but will have to be carried out eventually for all the nuclear plants.

In this context it is interesting to note that one of the conclusions of the Technology Foresight exercise carried out in the UK and published by the Office of Science and Technology in 1995 was that decommissioning, rather than new plant construction, was likely to be the main growth area for the UK nuclear industry (OST 1995). If, as some people would like, a full-scale nuclear phase-out programme was being considered rather than just a plant retirement programme then, depending on the time scale, there would be an even larger decommissioning task.

A nuclear phase-out?

Decommissioning may be difficult and expensive, but this is only part of the problem facing those who argue for a nuclear phase-out. According to some of the proponents of nuclear power, without an expanding nuclear programme the world would be faced with serious energy shortages. Fred Hoyle put this case forcefully in his classic book *Energy or Extinction: The Case for Nuclear Energy* (Hoyle 1980) and the idea of an 'energy gap' or shortfall has resurfaced regularly in energy policy debates in the UK and elsewhere.

While, as we have seen, it is conceivable that nuclear power could expand, the scale of its potential contribution in the short to medium term would seem likely to be limited. There are many other options which might be considered to be more credible if there was indeed an energy shortfall at some point. To some extent the prediction of a shortfall simply reflects the current lack of commitment to energy conservation. If the bulk of the world's growing population continues to use energy inefficiently then there could be serious problems as demand continued to rise dramatically. However, given a serious commitment to energy conservation around the world, the need for new supplies of whatever sort would obviously be lessened.

In subsequent chapters we will be looking at the overall global energy future and at the prospects for establishing a sustainable energy supply and demand system without the use of nuclear power, and without undermining economic growth. As we shall see, this does seem to be possible in technical terms, even with growing world demand. Indeed, given time and a major commitment to energy conservation and renewable energy development, it would also seem technically possible, in principle, to phase out the use of fossil fuel, a much more radical proposition.

However, opinions differ as to the way sustainable energy technology should be developed. For example, not everyone in the environmental movement feels that grand scale technical fixes like this can or should be used to try to sustain what they see as fundamentally unsustainable patterns of material consumption. That of course would make them even less sympathetic to nuclear power. Either way, for most environmentalists nuclear power has no place either in the short or long term and a rapid phase-out is often seen as urgently needed.

A shift away from nuclear power could present difficulties for some countries, but in general, it would seem that such a move should present few major practical problems in terms of energy supply. In most countries which have nuclear programmes there is actually an oversupply of electricity generation capacity. In the UK, for example, given the privatisation of the electricity supply industry and the so-called 'dash for gas', this overcapacity seems likely to increase. As and when renewables are deployed and energy conservation programmes introduced, there would be even less need for the nuclear element. By around the year 2015 to 2020 renewables seem likely to overtake the nuclear contribution in the UK and, given normal plant retirements, soon after then, only one nuclear plant would be left operating, without there being any formal commitment to phasing out nuclear power.

However, most environmentalists in the UK would prefer a more rapid and conscious phase-out plan, coupled with a commitment to renewables and energy conservation, a programme which they feel would also make sense around the world. As has been indicated, some countries are already exploring this option. The notable exceptions are France and Japan. France obtains about 75 per cent of its electricity from nuclear power plants and Japan around 50 per cent. However, as we shall see, Japan does have a substantial renewable energy development programme and, belatedly, France is embarking on renewable energy development.

The ex-Soviet nuclear programme, which provides around 12 per cent of the electricity used by the CIS, presents special problems. The safety of some of the reactors worries many observers, but few of the new Eastern and Central European states can afford to clean them up or to shut them down. Massive aid from the West would be needed to phase out nuclear power and stimulate renewables and conservation, the latter being vital given the very poor level of energy efficiency which exists at present. Some aid has been forthcoming but in the main the Western nuclear industry seems instead to be looking further eastwards to Korea, China and the Pacific rim as a new market for its wares. It is still something of a long shot, but far from a nuclear phase-out, we may yet witness nuclear expansion.

A nuclear future?

This is not the place to rehearse yet again the long running pro and anti nuclear debate. There are countless books, reports and tracts on both sides of the argument, some of which are mentioned in the Further Readings section at the end of the chapter. However, to round off the discussion it may be worth noting that the use of nuclear power presents a range of problems in addition to those already considered, which might become particularly important if and when nuclear power was taken up on a significant scale by the developing world. For example, there is the issue of the role of nuclear power in terms of economic development. Some developing countries evidently see nuclear power as part of the industrialisation process. High technology developments like this may benefit the technical and economic elites in some developing countries, but importing capital intensive nuclear power technology does not seem the best choice for those Third World countries which are struggling with large foreign debts, or whose populations in the main need cheap, simple, locally accessible sources of power. There are also concerns that part of the attraction of 'going nuclear' is that it can provide a means of developing nuclear weapons. There are other routes to obtaining the necessary nuclear materials, but civil nuclear programmes provide one potential source.

Although most, but not all, countries around the world have signed nuclear non-proliferation agreements, nuclear expansion raises concerns about the problem of illegal diversion of nuclear weapons-making materials like plutonium. These materials are carefully controlled but a black market has grown up, particularly since the collapse of the Soviet regime. Adding to the security problem by building more reactors would seem to be unwise.

New nuclear developments

Despite these arguments and the adverse economic assessments and spate of accidents, the nuclear industry remains a major player on the world energy scene. Given the poor prospects in their domestic markets, the Western nuclear companies see export orders as the key to the future. However, there have also been attempts to develop new technologies which might be seen to be targeted in part at the Western market. For example, in an attempt to assuage public concerns over safety, the industry has talked of developing so-called 'safe integral' or 'passive' reactor technology, using small modular units which are designed to be fail-safe (e.g. in terms of emergency cooling) under all conditions.

In the event, the emphasis seems to have moved instead on to larger conventional reactors, like the double-sized pressurised water reactor proposed as a follow-up to Sizewell B PWR on the Suffolk coast in the UK. The aim was to get unit costs down by opting for economies of scale with a very large plant, in the hope of getting back into the economic race. But there seems little chance of this happening, at least in the West, given that natural gas fuelled combined cycle power stations are so much cheaper. As already noted, in 1995 British Energy, the new UK nuclear company, announced that it was withdrawing plans for building further nuclear plants, including the double-sized Sizewell PWR.

With the economic case for nuclear power being less than convincing, the nuclear industry has made much of strategic arguments concerning energy diversity – the need to have a variety of generation options available. This might be more credible if nuclear power, both fission and fusion, had not soaked up the major share of research funding in the energy field for many decades. As it is, it could be argued that nuclear power has had its fair share of resources. Now it is the turn of renewables to show what they can do, especially since there is a wide range of renewable technologies offering what might be seen as more genuine diversity (Stirling 1994).

Summary points

- Nuclear fission produces radioactive materials that can be very hazardous to human health and which must be stored for many centuries.
- Reserves of uranium might be used up if an attempt was made to respond to global warming by building more conventional 'burner' reactors.
- Fast breeder reactors could extend the fissile fuel reserves, but at unknown cost, and the waste and plutonium proliferation problem would be increased.
- Nuclear fusion is a technological long shot with its own safety and economic problems.
- Concerns about safety, security and economics have led to a decline in enthusiasm for the nuclear option in some, but not all, countries.
- Nuclear power has had the major share of funding so far in the energy R&D field, but it might be argued that renewable energy should now be given a chance to show what it can do.

Further reading

Walter Patterson's much reprinted *Nuclear Power* (1976, Penguin, London) remains one of the best general introductions. There are many books and journals on specific aspects of nuclear power, ranging from the technical to the political.

Useful general sources from the pro-nuclear perspective are: P. M. S Jones, *Energy and the Need for Nuclear Power* (1989, Belmont Press, London); John Collier's *Nuclear Power: Clean Energy for the 21st Century* (1994, Taylor Blackhorn, London); the journal *ATOM*, published by the UK Atomic Energy Authority (AEA Technology as it is now known). The British Nuclear Industry Forum (BNIF) produces a range of materials on nuclear power, as do British Nuclear Fuels. On fusion, see R. Herman's *Fusion: Search for Endless Energy* (1993, Cambridge University Press, Cambridge) See also BNIF's *Fission, Fusion and Safety* (1994, Taylor Blackhorn, London).

Useful general sources from an anti-nuclear perspective are the two-volume compilation of papers and articles produced in 1991 by *The Ecologist* as a special report, 'Nuclear power: shut it down'; Peter Bunyard's *Nuclear Power:*

Way Forward or Cul-de-sac? (1992, *The Ecologist*); and the journal *Safe Energy*, produced by Friends of the Earth Scotland. Friends of the Earth and Greenpeace have produced a variety of publications on nuclear issues, including submissions to the various public inquiries on nuclear plant construction proposals.

Over the years there have been many popular books on nuclear issues, mostly from an anti-nuclear perspective. For example see Crispin Aubrey's *Meltdown: the Collapse of the Nuclear Dream* (1991, Collins & Brown, London). For a more scholarly analysis see Steve Thomas, *The Realities of Nuclear Power* (1988, Cambridge University Press, Cambridge); and the various studies on nuclear issues in the UK journal *Energy Policy*, for example, John Cheshire's 'Why nuclear power failed the market test', August 1992: 744–54.

For an international overview see P. R. Mountfield, *World Nuclear Power* (1991, Routledge, London). In addition, WISE, the Amsterdam-based World Information Service on Energy, is a major source of independent critical information on nuclear developments around the world (see Appendix II).

6 Renewable energy

- **Solar power**
- **Wind power**
- **Wave power**
- **Tidal power**
- **Wastes and energy crops**
- **Hydroelectricity**
- **Geothermal energy**

Natural energy flows and sources such as sunlight, the winds, waves and tides offer a relatively clean, safe and above all sustainable source of power. This chapter explores the renewable energy options, looking at the basic technology and at some current developments, and then reviews the overall global resource potential of renewable energy.

Renewable energy technology

Renewable energy is so called because it relies on natural energy flows and sources in the environment which, since they are continuously replenished, will never run out. So they meet our first criterion for sustainability. In what follows, to provide a context, we first look at the basic technology before moving on in Chapter 7 to look at developments around the world.

The basic technology is very varied, reflecting the range of natural energy sources. As has been indicated, most sources of renewable energy are the result, directly or indirectly, of the impact of solar radiation on the planetary ecosystem – the exceptions being tidal energy flows and geothermal heat. Incoming solar radiation provides energy for plant and animal life and drives the hydrological weather cycle – evaporating sea water and creating rain which feeds into rivers and streams. Local differential heating of the atmosphere, sea and land also causes winds, which move over the seas creating waves. By contrast tidal energy could be thought of as lunar power, given that the main component is due to the effect of the gravitational pull of the moon on the seas, although the sun's gravitational pull also plays a part. Geothermal energy is the result of residual heat deep in the planet, topped up by radioactive decay processes.

Renewable energy is actually already in widespread use: around 20 per cent of the world's electricity already comes from conventional hydroelectric dams, and in many countries wood and animal dung provide the only sources of power for cooking and heating.

Now however, a whole new range of renewable energy technologies is emerging. The sections which follow explore some of the key new technologies for harvesting renewable energy, looking first at solar power, including photovoltaic solar cells. We then move on to look at the basic physics and technology associated with the use of wind, wave and tidal energy, followed by energy crops, focusing on short rotation coppicing.

Solar power

The sun provides the basis for life on earth and delivers sufficient energy to each square foot to meet all our needs – if that energy can be tapped efficiently. Historically human beings have tried to do this via agriculture, using wood as a fuel and by harnessing the indirect solar energy represented by winds and streams. More recently use has been made of the stored solar energy of fossil fuels – coal, oil and gas.

However, the sun's heat can also be used more directly. In recent years there have been large-scale experiments with solar power – for example, via giant solar heat concentrating mirrors and dishes, tracking the sun across the sky and focusing its rays so as to raise steam for electricity generation. Large-scale 'solar thermal' plants like this are becoming increasingly popular in desert areas of the USA and elsewhere.

Solar energy can also be utilised on a smaller scale, via roof-top solar collectors (see Figure 6.1) which are in widespread use in many countries around the world, for example, in the Mediterranean region. Even in the UK, flat plate solar heat collectors, looking something like radiators mounted on roof tops and plugged into a hot water system, can typically cut average yearly domestic fuel bills by half; perhaps surprisingly there are around 300 solar houses in the London area alone (Kinsella and Bond 1985). However, in most UK locations the overall economics of solar space and water heating are currently not attractive compared with cheap

Figure 6.1 ***Flat plate solar collector for space and/or water heating***

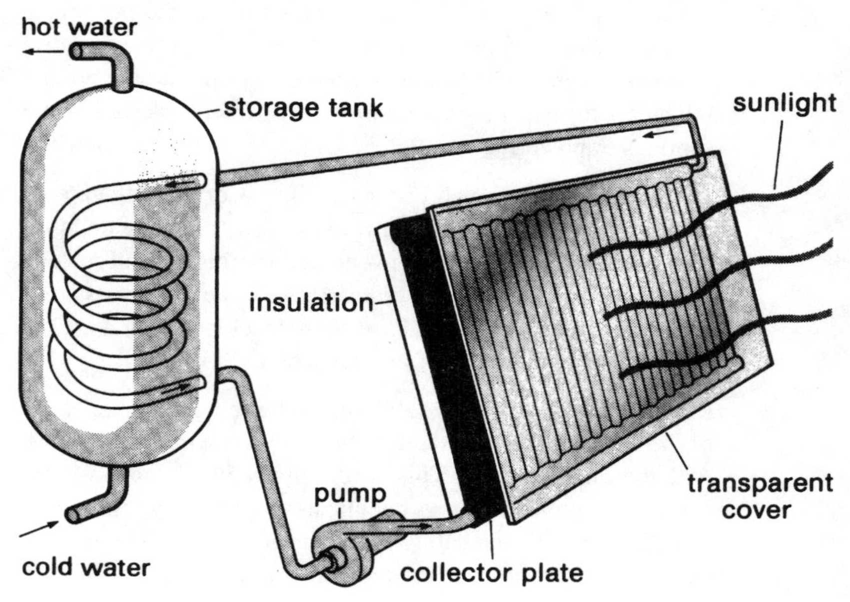

Source: Open University T362.

gas heating: commercial solar energy systems have payback times of five to ten years or more.

An alternative is the more cost effective **passive solar** concept. Conventional roof-top solar collectors use small pumps to drive the heated water around the heat circuit, but useful amounts of heat can also be collected by large south-facing glazed areas. This is much like the greenhouse concept – and involves no moving parts. Hence the term 'passive' as opposed to conventional 'active' solar collectors which involve pumps.

Typically, with a well-insulated house overall annual fuel bills can be cut by a third in this way. Large-scale solar atriums for offices and commercial buildings are now a familiar sight and at the more lowly level many people have added solar heat trapping conservatories to their homes.

However, the big breakthrough is likely to be in the **photovoltaic** solar field. Photo cells, like those on cameras and pocket calculators, convert sunlight directly into electricity. The only problem is that they are expensive. They were initially used mainly for powering space satellites, but developments in the semiconductor field have gradually brought down prices. Within a decade or so they are likely to be competitive with conventional power sources.

In this case, photovoltaic (pv) cells are likely to be used widely – even for domestic supplies. They are already to be found in some outlying rural areas where there is no grid electricity and the alternative is diesel, or nothing – in the Australian outback, in desert areas and in African villages – for water pumping, running refrigerators for key medical supplies or powering remote telecommunications equipment.

Soon there may be much wider scale use. Currently there is interest is using pv cells to substitute for roof or fascia cladding on buildings, thus saving on the cost of conventional roofing or wall materials and offsetting the cost of the pv cells. At the same time costs are dropping; new types of cell materials have been developed which are more efficient and cheaper and the use of pv seems bound to spread.

Environmentally, there would seem to be few problems, at least in terms of the use of pv cells, although there are some question marks associated with their manufacture. This is very much a high tech industry using exotic and often hazardous chemicals – potentially representing a significant health and safety problem for workers. There could also be ex-plant pollution issues to contend with. However, assuming these potential problems can be avoided, pv could become a key new renewable resource.

Wind power

In contrast to pv, wind power is already a significant energy source. The winds are an indirect form of solar power and they have been used for centuries as a source of energy. More recently wind power has become one of the more successful renewable energy technologies. The USA took the lead and California's wind farms now supply the energy equivalent of the domestic electricity requirement of a city like San Francisco. Europe has followed, and wind projects now exist in many other parts of the world; by 1995 the total generating capacity around the world was around 4,500 MW.

The implications of the basic physics of wind turbines can be relatively easily appreciated. The energy collected from a wind turbine is proportional to the area in the circle swept by the blade, i.e. Πr^2 where r is the blade length. The result of this 'square law' is that, for example, doubling the size of the wind turbine quadruples the power output, meaning that large turbines are much more effective. But beyond a certain size the blades face stress (and fabrication cost) limits. Currently the most cost-effective size is between 250 to 600 kW rated capacity (i.e. at full power) with blade diameters of between 25 and 45 metres.

The power in the wind is proportional to V^3, the *cube* of V, the wind speed. (The kinetic energy is $\frac{1}{2} mV^2$, where m is the mass of the air intercepted, which in turn is proportionate to V times the area swept.) So even a slightly higher wind speed site pays dramatic dividends in terms of available power.

However, the actual amount of power wind turbines can produce is less than the power in the wind flow they intercept. In part this is a matter of design and operational factors, but fundamental aerodynamic losses mean that the maximum amount of power that can theoretically be extracted is just under 60 per cent of that in the flow; there are also inevitably mechanical and electrical losses in conversion.

Wind turbines are often grouped together in 'wind farms' so that connections to the power grid can be shared, as well as control systems and road access for maintenance. Typically a separation of between 5 and 15 blade diameters is needed between individual wind turbines, to prevent turbulent interactions in wind farm arrays. This means that wind farms can take up quite a lot of space, even though the machines themselves only take up a small fraction of it, and this has led to some objections. It is argued that there would be insufficient room in countries like the UK to generate significant amounts of power.

It is relatively easy to work out the rough land area requirements for a wind farm, and then to calculate how much power might be generated in a country like the UK.

As Box 4 shows, even given the intermittancy of the wind, it would seem possible to replace 10 per cent of the UK's conventional power generation capacity with medium size 300 kW wind turbines by using between 1.4 per cent and 1.7 per cent of the UK's total land area, depending on the assumptions used. Obviously this is only a very rough estimate, based on some broad assumptions. The actual power output in practice would depend on the sites, and the operational patterns and more power might be obtained from wind turbines covering less area. Using larger machines or using array separations of less than 10 diameters would have the same effect. Equally, however, since wind turbines are visually intrusive, there will be specific siting constraints that could reduce net power availability: some high wind speed sites may not be acceptable. This issue will be explored in detail later.

Overall though, it would seem possible, even in a crowded country like the UK, to obtain around 10 per cent of the country's current electricity

Box 4

Power produced and land used by wind farms – a rough calculation

We can calculate the power output and area covered by a typical windfarm by making some very broad assumptions, using round figures, as follows. In very rough terms, a 10 by 10 array of 100 relatively small 300 kW machines (30 MW in all) each with say 30-metre diameter blades, and with 10-diameter separations, would cover an area of 3 km by 3 km. However only about 1 per cent of this area (i.e. 90 m^2) would actually be occupied by the base of the turbine towers, the rest could be used for agricultural purposes.

One hundred wind farms of this sort would give you 3 gigawatts of installed capacity – equivalent to three conventional 1 GW power plants.

Of course these wind turbines will not be able to operate continuously at full power since the wind is intermittent. Typically, given the variability of the wind, turbines in the UK can only deliver power on average for about 30 per cent of the time: this figure is known as the 'load factor'. The strict definition of load factor is actually more complex, but this will suffice for our purposes. However, it is worth noting that 30 per cent is only an average figure. The load factor figure for specific machines will depend on the wind turbine design, the site and the wind regime: higher load factors, up to 40 per cent and more, have been reported for modern machines on good sites, for example, in Scotland and the USA (Gipe 1995).

To make a fair comparison with conventional plants, it is important to realise that although fossil or nuclear fuelled plants do not have intermittent energy inputs and therefore have much higher load factors, they can still only achieve typical load factors of around 60 to 70 per cent. For the sake of simplicity let us use a 30 per cent load factor for wind turbines and 60 per cent for conventional plants. On this basis, to generate the same amount of power you would need twice as much wind farm generating capacity as you would conventional capacity. Put the other way round, you would expect wind turbines to generate about half as much continuous power as conventional plants with the same capacity. In the UK wind turbines and other renewable energy devices using intermittent sources are given a rating in terms of their **declared net capacity** (DNC) to reflect this comparison.

The DNC conversion figure adopted by the UK Department of Trade and Industry for wind turbines is actually 43 per cent, rather than 50 per cent, using the 30 per cent and 70 per cent figures (i.e. 30/70 × 100 per cent). So the actual equivalent DNC generation capacity of a wind turbine is considered to be 43 per cent of its full rated capacity. But for the sake of easy calculation, and given the fact that load factors seem to be improving, let us stick for the moment with a DNC conversion factor of 50 per cent.

On this basis, although our hypothetical 100 30 MW wind farms would in principle have a generating capacity of 3 GW, in practice they would be equivalent to only 1.5 conventional 1 GW plants (i.e. they would have a declared net capacity of 1.5 GW). In reality larger numbers of smaller wind farms would probably be used, although there is already one 103-turbine wind farm in operation in the UK. Also our calculation assumes relatively small 300 kW machines, whereas 500 kW machines are now common, and even larger devices are on the market. Obviously, although the separation distances between larger bladed machines would have to be greater, you would need fewer machines and less overall area for any given generation capacity.

Even so, wind farms would take up a relatively large area and in some countries this might be a problem. In the UK, for example, population densities are high and finding space for wind farms can be a problem. This issue has to be put into perspective. The UK currently has 60 GW or so of conventional installed capacity, leaving aside existing renewable sources like hydro and the combined heat and power element, so if 10 per cent of this was to be replaced by wind turbines we would require around 400 wind farms of the scale outlined above (i.e. with 100 300 kW machines). The tower area covered would be 36 km^2, while the total area covered by the complete arrays would be 3,600 km^2. That is just 1.44 per cent of the UK's total land area (250,000 km^2). If we used the 43 per cent DNC figure, this would mean that we would need about 465 wind farms, but they would still only cover 1.67 per cent of the UK's total land area.

Using larger machines would of course reduce this, although these may be felt to be more visually intrusive. However, it should be remembered in this context that, because of the 'square' law linking power and blade size, doubling the generation capacity of the turbine does not require doubling the blade diameter, so larger capacity machines do not need to be proportionately larger physically.

requirements by using, say, 2 per cent of the land area. That, as it happens, is roughly the area that has been estimated as likely to be suitable and available for wind farm projects in the UK without significant intrusion (Clarke 1988).

The offshore potential is much less constrained. For example, according to the UK government's Energy Technology Support Unit, between 100 and 150 TWh p.a. could in theory be obtained from sites in shallow water off East Anglia (i.e. up to 50 per cent of mid-1990s UK electricity requirements).

Offshore wind projects are already being developed by Denmark, Germany and Sweden. Although there are extra costs associated with offshore siting and power transmission by marine cable back to shore, these costs are at least partly offset by the generally higher speed and more consistent winds offshore. As acceptable sites for land siting become harder to find, offshore wind could well become a major energy option.

Wave power

Wave power represents the other major offshore energy resource. Waves are caused by the frictional effect of wind moving across the open sea – a series of rolling circular motions being set up under the surface. In effect waves are a form of stored wind energy. Typical energy densities at good sites (e.g. 100 miles or so off the north west of Scotland) are on average around 50 kW per metre of wave front.

A series of floating wave energy converters, made up of large individual units linked together in chains, could be used to absorb some of this energy. Given a series of chains of total length 400 km, with a conversion efficiency of 30 per cent, the generating capacity would be 6 GW. A scheme of this scale would be theoretically possible off the North Atlantic coast of the UK and, if built, it would be equivalent to about 10 per cent of the UK's total current installed generation capacity. Indeed, in principle you could have larger schemes.

The actual total wave energy potential would obviously depend on how many chains of converters were installed, their overall conversion and transmission efficiencies and the location of the devices. Estimates for the UK range from 20 per cent of current UK electricity requirements up to 50 per cent or more, using sites off the north west and east of Scotland and to the west of Cornwall (Ross 1995).

Given its Atlantic positioning, the UK has most of the best sites and about a third of the total European wave energy resource. Even so there are significant resources elsewhere in Europe, for example, off Ireland, Norway, Portugal and Spain. Similar resources exist elsewhere, for example, off Japan and Australia.

Scale prototypes of some of the conversion systems have already been tested in the UK, perhaps the simplest being a squeezed air bag system (the Clam) – see Figure 6.2. The most complex is Salters gyro stabilised 'nodding duck', a scale model version of which has been tested in Loch Ness. Clearly, when operated out at sea they would have to be engineered to withstand storms, but maintenance problems could be eased by towing individual units back to shore; the offshore position would attract few environmental siting constraints.

Figure 6.2 ***SEA-Lanchester wave energy device***

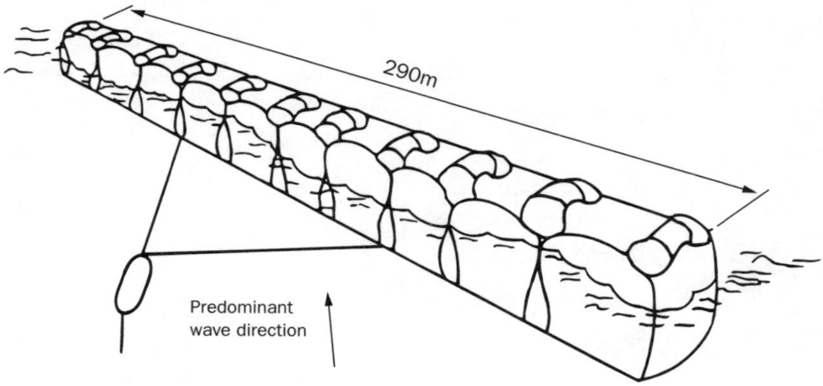

290m

Predominant
wave direction

Source: SEA-Lanchester, Coventry University.

Smaller amounts of power, at possibly less cost, could be obtained from devices operated nearer to the shore ('inshore' or 'coastal' systems) and from shore-mounted units, e.g. sited in gullies ('onshore' systems). One small 75 kW onshore system has been constructed at Islay in Scotland and a 2 MW inshore device, the OSPREY, is being developed (see Figure 6.3). However, this has met with some problems: during its initial test in 1995 the first prototype was damaged by storms. The developers are pressing ahead with a second version. Many other inshore and onshore devices have also been developed elsewhere, notably in Norway and Japan.

Although the total amount of power from inshore and onshore systems may be relatively small, the costs are lower than those for the more complex deep sea wave energy systems. Small-scale wave power systems of this sort are seen as being likely to be of particular relevance to developing countries with suitable coastlines and to remote islands where the only source of power might otherwise be imported fuel.

Tidal power

The tides are the result of the gravitational pull of the moon modified by that of the sun, the tidal height sometimes being increased by local funnelling effects in estuaries. The power that can be generated by a tidal barrage across a suitable estuary can be calculated in rough terms by assuming that the barrage traps a constant area A of water at high tide, which is then passed through turbines, to the low tide level, falling through a distance r, the high to low tide range. The mass of water is dAr,

Figure 6.3 ***ART's 2MW OSPREY-1 Ocean Swelled Powered Renewable Energy device, designed for installation on the sea bed in shallow water 300m offshore***

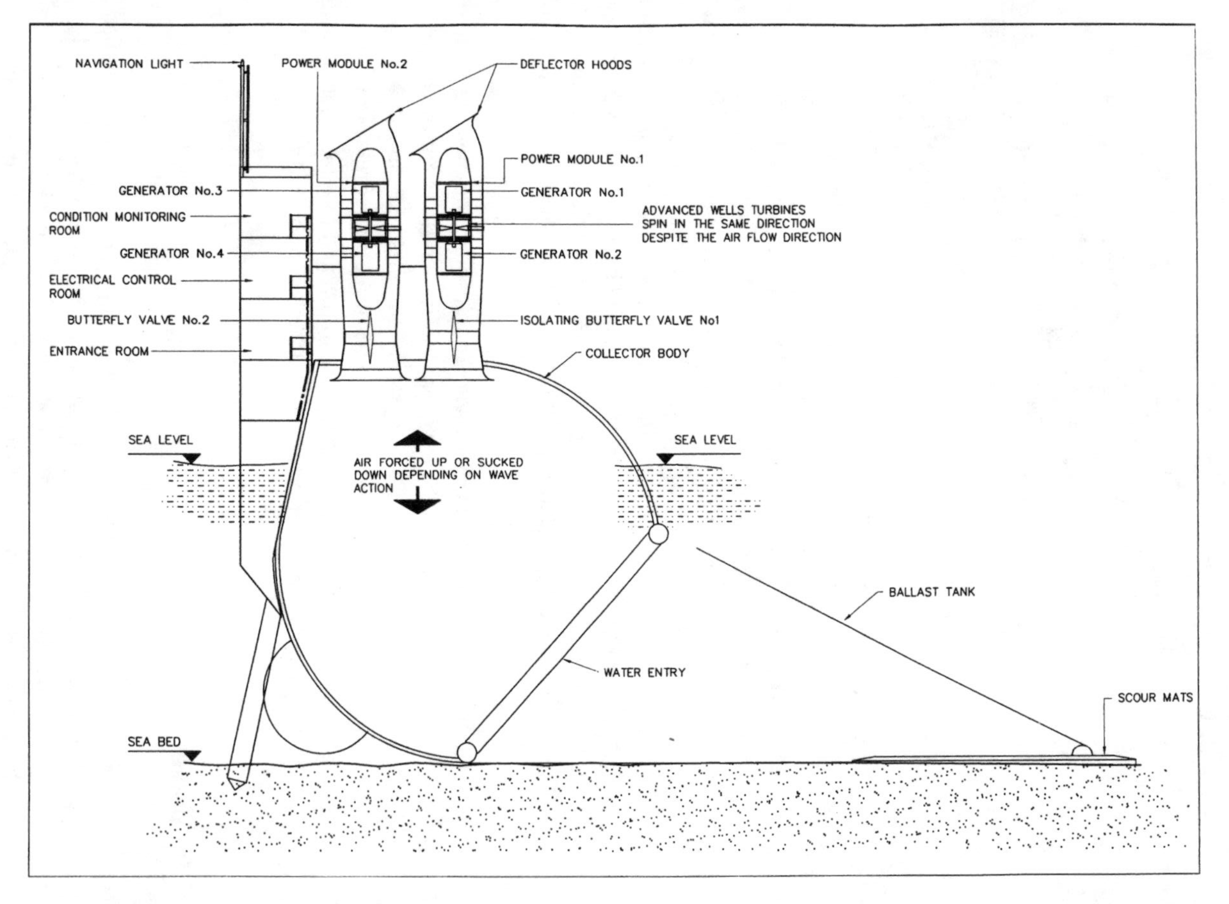

Source: Applied Research and Technology, Scotland.

where d is the density of water, and it would fall on average through $\frac{1}{2}r$. The potential energy per tide is thus $\frac{1}{2}dAr^2$.

This square law implies that estuaries with large ranges will pay dividends. Hence the interest in the Severn estuary in the UK where topographic funnelling effects produce tidal range of up to 11 metres.

Since there are only two tides per day a tidal barrage will not operate continuously – typically they can supply power for say three to five hours per tide, depending on the operational pattern adopted (e.g. generating on the incoming flow, on the ebb or on both, using two-way turbines).

The proposed 11-mile long 8.6 GW ebb generation Severn tidal barrage (see Figure 6.4) would, for example, generate 17 TWh per annum in total, around 6 per cent of annual UK electricity requirements, although the value of this would vary since not all of this would be available at times of peak demand.

This disadvantage can be offset to some extent by using barrages as a short-term pumped storage reservoir (e.g. using off-peak electricity from other plants to pump water behind the barrage ready for electricity generation via the barrage turbines when needed). In addition, if there are several barrages in different parts of a country (e.g. in the UK on the Humber, Mersey, Dee, Solway Firth, etc.) the net output would be more nearly continuous, since the tides at each point will occur at different times, the difference being up to five hours for some UK sites (Watson 1994).

Figure 6.4 ***Artist's impression of the Severn tidal barrage concept; the 11-mile long barrage would have 8GW of turbine capacity***

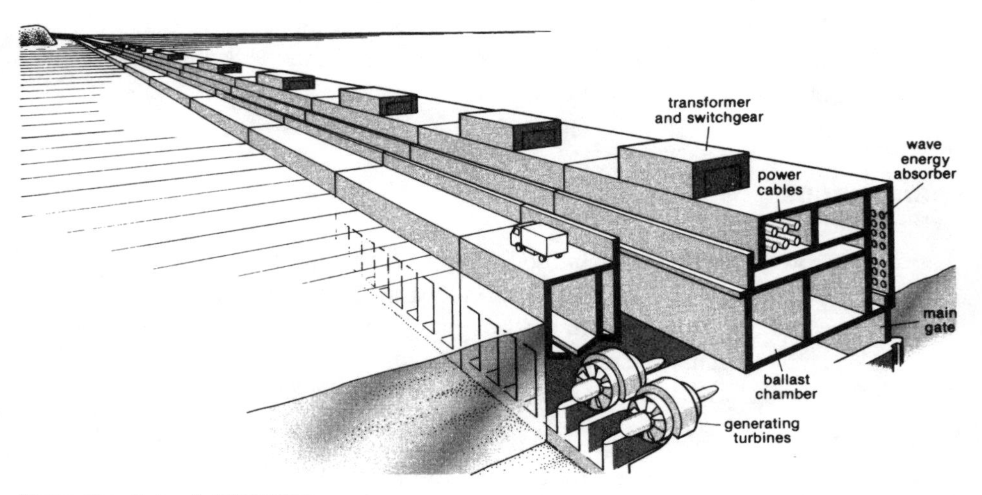

Source: Open University T362/T302.

The basic technology of power generation is similar to that used in **hydroelectric dams** – large turbines are mounted in vast concrete structures – although the head of water developed by tidal barrages is much less than that in conventional hydro projects.

However, with tidal barrages there is no need to impound and flood vast new areas – this being one of the reasons why objections have increasingly emerged in relation to large hydro schemes. Nevertheless, proposals for tidal barrages have met with significant objections from environmentalists concerned about the likely negative impacts on the local ecosystem of altering the tidal range. Extensive studies have already been carried out on the Severn estuary, but much careful environmental impact assessment work still needs to be done to resolve this issue.

Although the UK has an enviable tidal energy resource, put at around 20 per cent of electricity requirements if fully developed, there are also opportunities for tidal energy elsewhere in the world. A 250 MW tidal barrage already exists on the Rance estuary in Brittany and there are smaller barrages in the former Soviet Union, Canada and China. The total world tidal barrage potential is put at about 120 GW, which could produce around 190 Twh p.a. (Baker 1991).

Rather than having to build expensive barrages, there is also the option of collecting energy using submerged wind turbine-like devices from the fast moving **tidal currents** that exist in channels offshore in some areas. Some prototypes have already been developed including a 10 kW device tested in a loch in Scotland in 1994 (see Figure 6.5). According to a study by ETSU, in principle the UK might obtain up to 19 per cent of its electricity if all its tidal stream resources were tapped (ETSU 1993a). Similar resources exist elsewhere, for example, in the Straits of Messina between Sicily and mainland Italy.

Energy crops

Growing fuel rather than food could well become a significant new option for the world's farmers. As long as the replanting rate matches the rate of use, the overall process of energy crop growing and combustion can be 'greenhouse neutral' in that plants absorb carbon dioxide while growing, so that no net carbon dioxide is produced.

Interest has already be shown in 'biofuels' of various sorts – liquids (like ethanol and biodiesel) for transport use, and gases (like methane), or solids (e.g. wood) used for heating or electricity production. Ethanol has

Figure 6.5 ***IT Power's tidemill concept for harnessing tidal currents***

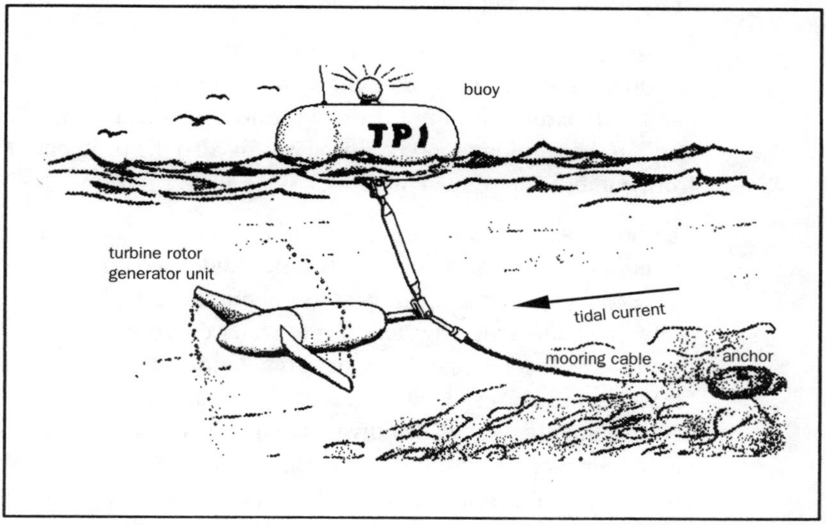

Source: IT Power.

been produced from sugar cane in Brazil for some years. Biodiesel (rape methyl ester) is now being produced in some parts of the EU, especially in France, from oil seed rape. However there are some questions as to the net energy ratio (energy input compared to energy output) of the total growing/harvesting and transport system.

Solid fuels from 'woody biomass' are another option: the energy ratio is believed to be higher. **Short rotation arable coppicing**, for example, using fast growing willows or poplar, is currently seen as likely to be an important source of fuel for UK electricity generation. Indeed the UK government's Department of Trade and Industry (DTI) has estimated that the maximum total realistic resource potential by the year 2025 could be up to 150 TWh/yr – half current UK electricity requirements (DTI 1994b).

Short rotation coppicing (SRC) does not involve full tree development – instead the thin willowy growths, reaching perhaps fifteen feet, are cut back to stumps every three to five years. Large areas could be involved with mechanised cutters passing periodically along corridors through the coppice plantations. The resultant wood chips would be transported periodically to combustion plants. The combustion process can be very efficient, especially given the development of advanced co-generation/CHP techniques. For example, with gasified biomass used to power steam-injected turbines (the so-called BIG–STIG technology), operated in the combined heat and power mode, conversion efficiencies of up to 80 per cent can be obtained.

However, there remain some environmental concerns, for example, in relation to the run-off of pesticides and herbicides, the impact on wildlife and local disruption from the extra traffic movements required for moving the resultant wood chips. There is also the problem of emissions from combustion of the chips.

The proponents of SRC argue that, in fact, herbicides would only be used in the first year of the cycle and less pesticides would be needed than for conventional arable crops. They also claim that coppices tend to mop up pollutants, thus reducing run off into water courses. Coppices would, they say, be located mainly on marginal land and can offer habitats for birds and insects, especially given the open access corridors that would traverse the plantations. Biodiversity can be ensured by using multi-cloned plants or by mixing species: coppices have already been shown to attract woodland animal species. Overall, if degraded or abandoned cropland were used, the environmental and wildlife impact would be positive.

The proponents of SRC argue that there would not be excessive transport requirements. Wood chips would be stored on site at farms and transported annually to combustion units on local industrial sites. For a coppice plantation with 5 hectares harvested annually, only 75 tonnes of wood chip would need transporting annually. This would involve 4 × 20 tonne loads or 8 × 10 tonne loads – not a huge annual volume and it would replace the transport of the crops which might otherwise have been grown.

On balance then arable coppicing seems to offer significant advantages. However, there are clearly some unknowns: what exactly are the environmental and pollution implications; how economic will it be; how will it be planned; what scale is appropriate? Is it just an extension of conventional farming or something new? At the very least, some form of planning control would seem necessary to avoid land use conflicts and visual disruption. Even so, energy crops could well become a major new energy source.

Biomass wastes

The idea of producing specially grown energy crops is relatively new, but the various types of biomass waste materials already produced in modern societies also represent a useful source of energy. For example, there is a whole range of farm wastes including pig slurry and chicken manure. Some animal and agricultural wastes can be converted into methane gas via anaerobic digestion techniques, while others can simply be burnt as a fuel; there are now power plants in the UK fuelled with chicken droppings. Residual wood from forestry operations can also be used as a fuel, as can straw.

In addition to these agricultural sources, there is a further source of biomass available in industrial societies in the form of domestic and industrial waste. **Energy-from-waste** projects of various types are already widely established around the world. Electricity generation using waste combustion is a well-developed option, although as projects have spread there have been concerns raised by environmentalists about toxic emissions. The collection of methane gas from landfill sites is another option. There can be problems with leaching out of toxic materials from landfill sites into the local watercourses, but since landfill sites exist in large numbers it would seem better to gather the gas they produce, rather than letting it leak out and contribute to global warming. This also reduces the risk of explosions at the landfill sites.

However, not everyone is convinced that waste should be seen as a genuinely 'natural' or renewable source of energy. Certainly large amounts of industrial and domestic wastes are produced regularly in industrial societies, but some environmentalists argue that the generation of material waste (for example, from packaging) should be minimised in the first place and that the energy content in whatever is left can be recovered more effectively by materials recycling schemes (Greenpeace 1992).

On this view, direct energy recovery via combustion or landfill gas should only be a last resort. The counter viewpoint is that recycling can be energy intensive and that, even with a major commitment to waste minimisation and recycling, there will still be a significant amount of waste, representing a large potential source of energy (Porteous 1992).

Other renewables

There are many other sources of renewable energy ranging from giant schemes for making use of the temperature differentials between the surface and the depths of the oceans (the so-called Ocean Thermal Energy Conversion technology or OTEC) to the use of **geothermal** heat deep underground. OTEC has yet to be developed seriously, but geothermal energy is being used widely around the world.

Figure 6.6 ***Artist's impression of a geothermal 'hot dry rock' well***

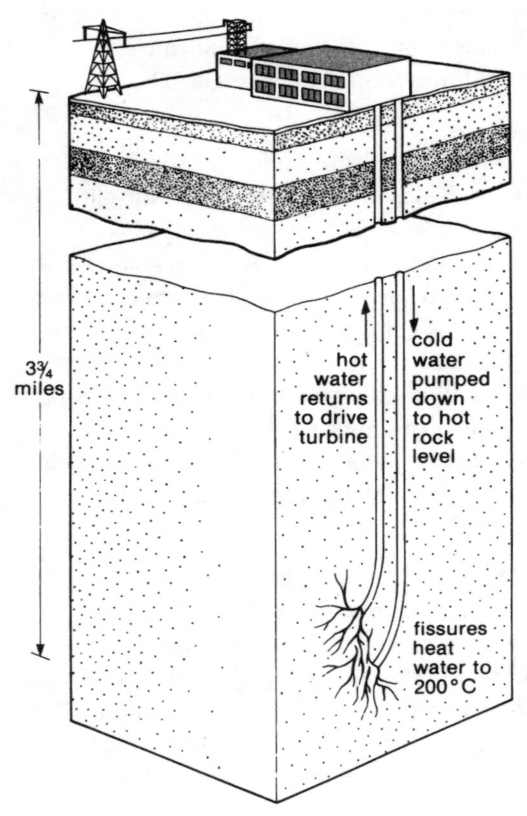

Source: Open University T362.

Extracting heat from geothermal sources deep underground is not strictly a continuously renewable process: the heat gradient is gradually diminished, although it may be replenished later if heat extraction is halted for some years. The energy comes from the heat deep in the core of the planet, which in part is the result of radioactive decay processes, so geothermal energy could be seen as 'natural' nuclear energy.

Natural hot water and steam geysers exist in various parts of the world, for example in the USA, and the UK has a number of hot 'spa water' sites. This type of geothermal heat comes from aquifers near to the surface. The resource is tapped for hot water or power production in many countries, e.g. in Japan and the USA. In addition to providing around 34 Twh of direct heat supplies, in 1994 there was a total of around 6.5 GW of geothermally powered electricity generating capacity in place around the world; by the year 2000 it could reach 10 GW (Fridleifsson 1996).

An even larger energy resource can be tapped if access can be gained to the so-called 'hot rocks' deep underground. The technology for deep well heat extraction from 'hot dry rocks' typically involves drilling two bore holes to around 3 to 4 miles depth in suitable locations and then forming a series of geological fractures in the rocks to connect up the bottom ends of the bore holes. Water is then injected under pressure down one bore hole so that it percolates through the fracture pattern, which acts as a heat exchanger, to re-emerge up the other bore hole as steam. This can then be used to drive turbines for power generation (see Figure 6.6).

Finally, there is **hydroelectric** power which has been exploited on a large scale for many years. This can also be harvested on a smaller scale – as in so-called 'micro-hydro' projects in rivers and streams. The individual generation capacities involved are small, but taken together they represent a significant amount of power. This option may be particularly relevant in developing countries, providing cheap local power inputs. Micro hydro also avoids the environmental problems associated with massive hydro schemes.

The total renewable resource

Having now reviewed some of the key technical options, what is the overall scale of the renewable energy resource? The simple message is that the basic energy source is very large. The sun delivers a vast amount of energy to the earth, much more than could ever be used, and the resultant natural energy flows represent a very large, non-finite resource base, with the solar radiation input alone amounting to around 90,000 terawatts. For comparison, in 1990 total global energy consumption, measured on a continuous equivalent power basis, was only 13.5 TW. Not all of this 90,000 TW can be successfully captured and used. Most of the flows are diffuse, some are intermittent and the efficiency of conversion technologies has to be taken into account, as does the location of the source.

Projections vary as to the amount of useful energy that can be extracted but Jackson, reviewing a series of papers in *Energy Policy* journal, put the world's recoverable tidal resource at 0.1 TW, the hydro resource at 1.5–2 TW, the wave resource at 0.5–1 TW, the wind energy resource at 10 TW and the solar resource at 1,000 TW (Jackson 1992).

Of course, what matters is the extent to which it can be developed in practical terms and the timescale on which it might be achieved. This will depend among other things on the success of the technological development programme, the economics and the environmental constraints and crucially, for a new area of technology, on the level of support given. We shall be looking at the state of play around the world in Chapter 7, after which we shall be discussing some of the strategic development issues.

Summary points

- Most renewable energy sources derive directly or indirectly from the sun, the exceptions being tidal power and geothermal energy.
- Of the 'new' renewables, wind power has been developed quite widely, as

has solar power. Wave and tidal energy are less widely used but energy crops are beginning to become established as a major new option.

- The renewable energy resource is very large, but only a small part of it can be tapped effectively and economically.
- Nevertheless, if fully developed, the use of renewable sources could, in principle, meet all the world's energy requirements.
- Renewables are still only a relatively marginal source, but their use is growing around the world.

Further reading

The Open University course reader *Renewable Energy: Power for a Sustainable Future*, G. Boyle (ed.) (1996, with Oxford University Press, Oxford) provides an excellent introduction to the technology, with the emphasis on the UK.

The classic (but massive) text is Bent Sorenson's *Renewable Energy* (1979, Academic Press, London). John Twidell and David Wier's *Renewable Energy Resources* (1990, E & F N Spon, London) is a shorter text providing a good introduction to the basic physics involved. For a less technical guide see Mike Flood's *Energy Without End* (1991, Friends of the Earth, London).

A useful US source is Chris Flavin's *Power Surge: Guide to the Coming Energy Revolution* (1995, Worldwatch Institute, Washington DC). See also Michael Brower's *Cool Energy: The Renewable Solutions to Global Warming* (1990, MIT Press, Cambridge MA in association with the Union of Concerned Scientists).

There are a number of useful books on specific technologies including Paul Gipe's *Windpower Comes of Age* (1995, Wiley, Chichester); Clive Baker's *Tidal Power* (1991, Peter Peregrinus/IEE, London); and *Power from the Waves* (1996, Oxford University Press, Oxford) by David Ross.

7 Renewables worldwide

- UK renewables
- US renewables
- European renewables
- Japanese renewables

Renewable energy technology has developed rapidly in recent years and it is being used widely around the world. This chapter explores the state of play so far, focusing on developments in the USA, Europe and Japan, and looks at the reasons why some countries are urgently pushing ahead with renewables.

Renewables old and new

The use of **renewable energy** is not new. As we noted earlier, biomass in the form of firewood still represents the main fuel source for many of the world's people and conventional large-scale hydroelectric generation is a major existing use of a renewable energy source. However, wood fuel is becoming increasingly scarce and given the high capital cost of large hydro projects and, in some cases, major environmental impacts of such schemes, there has been growing interest in smaller scale hydro plants around the world.

Of the 'new' renewables, the use of solar energy for space and water heating obviously has attractions in many countries. There are major projects underway in the USA and elsewhere for electricity production, either using solar heat to raise steam, or using sunlight to power photovoltaic cells. However, in terms of electricity production, it has been wind power that has made the biggest impact: by the mid-1990s there were more than 20,000 wind turbines installed around the world.

While the review of international developments that follows focuses on the UK, USA, continental Europe and Japan, it should be noted that renewable energy technology is also being deployed quite widely elsewhere. For example, India and China both have major wind power programmes, with 2,000 megawatt targets. In many developing countries the use of solar energy is an obvious option. Some Middle Eastern counties are interested in exploring the idea of hydrogen production powered by photovoltaics as a long-term alternative to reliance on oil sales. Saudi Arabia is collaborating with Germany on a 'Hysolar' pv project. For the moment, though, as this example illustrates, most of the projects are being set up in the 'developed' countries.

Developing renewables in the UK

As a hilly country surrounded by stormy seas, the UK has quite large renewable energy resources, possibly the largest per capita in the world. For example, the overall UK resource potential for electricity producing renewables, generating at a cost of less than 10 pence/kWh (and assuming an 8 per cent discount rate on investment) has been estimated by the UK government as being up to 1,100 TWh/yr by the year 2025 – more than three times the current (mid-1990s) 300 TWh annual national demand for electricity (DTI 1994b).

This of course is only the theoretical maximum. But even taking account of practical constraints (such as finding acceptable sites for wind farms) the UK clearly has a very significant renewable energy resource. This first became apparent when ETSU, the government's Energy Technology Support Unit, started to assess the renewable options.

ETSU was set up following the oil crisis in 1974 and oversaw and assessed a wide range of government-funded research and development projects in the renewable energy field. Wave power, tidal power and geothermal energy were initially the favoured options, but from around 1984 onwards windpower and, more recently, energy crops came to the fore. We shall be looking at some aspects of this programme in later chapters. By 1994 a cumulative total of around £232 million had been allocated.

However, since then the level of R&D support has begun to fall, reaching a peak of around £25m in 1994–5. The government argued that the R&D programme had successfully identified those options which were likely to be commercially viable – namely on-land wind, waste combustion, some energy crops and micro hydro (DTI 1994b). This strategy of focusing on the currently commercial options has its limits: the renewable energy field is very new and it is perhaps rather short-sighted to cut funding for the longer term projects (Elliott 1996).

The non-fossil fuel obligation

Nevertheless, the government has accepted that some interim support was needed for those renewable energy technologies that were near but had not quite reached the commercial lift-off point. In order to support the commercialisation process, in 1990 the government introduced a special interim cross-subsidy arrangement funded by a small levy (less than 1 per cent) on fossil fuel supplies, with the extra cost being passed on to electricity consumers. This levy was actually part of a larger scheme introduced primarily in order to support nuclear power. The so-called 'fossil fuel levy', of around 10 per cent, raised around £1.3 billion each year, most of which went to offset the cost of nuclear generation. The renewables only obtained a very small part; £30m p.a. in the first phase of the scheme, although this has gradually increased, and is expected to reach £150m by the year 2000, with the nuclear allocation being withdrawn as a consequence of the privatisation of the bulk of the UK's nuclear plants in 1996.

Buttressing the levy scheme, the government imposed a statutory obligation on the regional electricity companies which supply power to consumers, requiring them, under the terms of the government's **Non-Fossil Fuel Obligation** (NFFO) to buy in specified amounts of non-fossil derived electricity. The levy was provided in order to offset the extra cost of meeting this obligation.

The nuclear part of the NFFO was set at the entire output of the nuclear plants in England and Wales (around 8.5 GW); separate but similar arrangements were made in Scotland. In parallel a series of NFFO orders emerged specifying how much renewable energy capacity the regional electricity companies had to contract for. The first was in 1990 and set at 102 MW declared det capacity. (As you may recall from Chapter 6, Box 4, the declared net capacity is the capacity rating adjusted to take account of the fact that some renewable energy technologies make use of intermittent natural energy sources like the wind.) Subsequently further rounds of the renewable NFFO have emerged and similar schemes have been put in place in Scotland and Northern Ireland. The overall target for the renewable part of the NFFO has been set at 1,500 MW declared net capacity by the year 2000 (DTI 1994b).

This of course is only a fraction of the full potential of the UK's renewable resource. As has already been noted, the UK's overall renewable energy resource has been put at around 1,100 TWh/yr or more than three times current UK electricity consumption. This is of course the maximum theoretical potential, ignoring many practical economic and siting constraints. A more realistic, fairly conservative mid-range estimate, adopted by the government's Renewable Energy Advisory Group, is a 20 per cent contribution to UK electricity requirements by 2025, with around 10,000 MW (net) of renewable capacity generating about 60 TWh/yr (REAG 1992).

Even though the UK is only developing its renewable energy resource relatively slowly and has cut back on research, as ETSU's studies have shown, it clearly has an enviable renewable energy potential (ETSU 1994) with, for example, among the world's best wind, tidal and wave potential. However, many other countries also have major renewable energy resources and, as the next sections illustrate, renewable energy technologies are being employed on an increasing scale around the world, in some cases with more enthusiasm.

The US renewable energy programme

The USA has a quite significant renewable energy programme. In effect, the use of renewables, along with the use of natural gas, has taken over from the post-war US nuclear programme.

Following the 1974 oil crisis, large-scale federal government R&D funding was provided for renewable energy, rising to over $1.2 billion in 1979–80. Funding was cut back dramatically during the Reagan/Bush years but, with the Clinton administration and the advent of concern over global warming, it has subsequently grown steadily, from $119 million in 1990 to $240 million in 1992, rising to $274 million in 1994, with a $337 million programme being proposed for 1995, although, in 1996, the Republican majority in the US Congress imposed cutbacks.

The USA already had a large hydro capacity, but the development of the 'new' renewables brought the total renewable generating capacity to around 15 per cent of total US generation capacity. By 1991 there was a total of 85.47 GW of renewable capacity installed, rising to 102 GW if pumped hydro storage is included and possibly higher if the increasing numbers of off-grid independent projects are included (Anderson 1994).

Of the new renewables (i.e. apart from hydro) various types of energy crops have been developed most extensively, followed by geothermal (steam and hot water) and wind power. Wind technology is developing rapidly and is seen as competitive with conventional sources in some contexts. California's wind farms contain around 16,000 privately owned wind turbines. Wind farms are also being installed in several other states. Large-scale solar thermal (with focused sunlight used to raise steam for electricity generation) is increasingly seen as a viable option and photovoltaic solar systems are also gaining ground, especially for off-grid independent generation.

Support mechanisms

There is a range of financial support mechanisms for renewables in the USA. Some are quite complex and we can do no more than summarise the basic arrangements.

As indicated above the federal government has provided significant funds for R&D over the years, while some states have provided tax concessions and other forms of subsidy designed to stimulate the market.

More generally there is the Public Utilities Regulatory Policy Act (PURPA) introduced in 1978, which, in essence, provides premium rates for renewable energy projects. In addition, there has been a move in some states to introduce 'least cost planning' and the more advanced 'integrated resource planning' financial assessment techniques. These can be used to assess the relative costs and benefits of all the energy options, including conservation and renewables, before making investment choices. They can be used to take account of the relatively high environmental costs of conventional energy systems. Interestingly, the process of technology choice is open to local consumer and community involvement, via the work of the Public Utility Commissions (PUCs), state level regulatory bodies which hold public hearings. In the UK context, this would mean that each regional electricity company had to have public meetings on its technology investment programme.

As might be expected, this form of consultation does not always favour renewable energy projects which, in some contexts, may cost consumers more. Some PUCs have tended to push for gas as the cheapest option. More generally, there has been resistance to what some see as bureaucratic interference with free markets resulting from the PURPA and PUC schemes. Given the recent restructuring of the US electricity utilities and the parallel moves towards deregulation, the impetus behind integrated resource planning has been weakened. But environmental concern is still relatively high on the US political agenda, as witness the relatively strong emission control standards. Given the regulatory and legislative context in the USA, renewable energy projects can now sometimes be the most economically attractive new generation options. The end result is that some of the more progressive PUCs have been able to promote renewable energy projects. But whether this trend will continue in the new post-restructuring context remains to be seen.

Renewable energy in continental Europe

Continental Europe is seeing similar levels of expansion in renewable energy development with environmental concerns being perhaps even more of an impetus. The following is in no way a comprehensive review, but is rather an attempt to give a feel for the pattern of development.

Denmark has been a pioneer in this regard, in part due to its decision not to develop nuclear power, with wind power being its most obvious commitment. The financial support structure operated involves a range of subsidies for operators of wind turbines and other renewable systems. As a result of this, more than 3,500 wind turbines have been installed so far, about 70 per cent of them being owned by local people via local co-operative 'guilds' which by 1991 had installed around 300 MW of wind turbine capacity.

Overall, Denmark has set a target of 1,000 MW of wind power capacity by 2005, as part of its commitment to reducing greenhouse gas emissions by 20 per cent by that year. By 1995 the total wind capacity had reached 600 MW, around 4 per cent of Denmark's electricity requirements. Some of the existing wind plants are being replaced with larger more efficient units but, with land for new sites being at a premium, Denmark has also pioneered the world's first offshore wind farm, an 11-turbine 3 MW project at Vindeby. In addition to wind power, Denmark's Energy 2000 programme puts heavy emphasis on energy conservation and the efficient use of conventional fuels. Biomass, particularly straw, is increasingly being used as an alternative fuel for Denmark's extensive district heating networks.

Sweden has followed the Danish example, having decided to try to phase out its nuclear plants. Initially several large wind turbines were developed with some $40m allocated to windpower between 1991 and 1996. There are plans for a 98-turbine offshore wind farm. There is also a range of other renewable projects, most notably involving biofuels.

The **Netherlands** has a significant wind programme, with a target of 450 MW of wind capacity by the year 2000. Its financial support structure is similar to Denmark, with grants of up to 40 per cent of the capital cost of wind projects being typical. There is also an offshore wind project, 2 MW near Medemblik, and a 200 MW offshore wind target has been set. **Norway** already generates the bulk of its power from hydro and has also been developing wave energy systems, as has **Portugal**.

In general, the focus in each country is shaped by its geography and the renewable resource that it defines. Countries with North Sea coasts are interested in wave power and the northern European countries have good wind resources. By contrast the southern countries tend to focus on solar projects. For example, the EC has supported large scale solar thermal projects in **Italy** and active solar power is widely used in **Greece**. However, both countries are also interested in wind power. **Spain** also has a significant wind programme, with something of a US-style 'wind rush' underway and a target set of 750 MW by 2000.

Biofuels like rape methyl ester or 'biodiesel' have come to the fore in many European countries, particularly in **France**, where set aside land is used to grow this energy crop. **Austria** has also made extensive use of biomass for fuelling district heating networks.

Tidal power is another geographically determined technology. The UK has some excellent sites but so far Europe's only major tidal energy project is a 240 MW barrage on the Rance estuary in Brittany, completed in 1967 and providing something of a contrast to France's large nuclear programme. Interestingly, however, following a review of energy policy in 1995, France now seems to be making a larger commitment to renewables, particularly to wind power.

Germany, a relatively late entrant into the renewable energy field, has followed Denmark's example and is heavily supporting wind power, with high levels of national and local state support, via grants and market subsidies. The total allocation for the market development of wind power between 1979 and 1994 has been $75 million, while the German government allocated some DM328 million to R&D on wind technology between 1974 and 1993. Some 1,800 wind turbines had been installed by 1993, with around 334 MW of total installed capacity and by 1995 the total wind capacity had reached 1,100 MW – outstripping Denmark.

The scale of development and deployment of photovoltaic solar cells in Germany has also been quite significant. For example, the R&D budget for pv was more than DM100 million in 1991 and there is a DM80 million '1000 roof' photovoltaic solar module demonstration programme, with around 25,000 pv systems being installed on rooftops through the country.

Overall, Germany had some 5,393 MW of renewable capacity installed by 1992 (including the large existing hydro capacity) representing around 5 per cent of total generating capacity, with 'new' renewables (wind, biomass and photovoltaics) making up just over 1,000 MW of the total renewable capacity (Anderson 1995).

Since then wind and pv have been pushed even harder, with significant subsidies and clearly Germany sees renewables as being of major importance. It is not hard to see why. Germany imports over 55 per cent of its fuel and the expansion of nuclear power has been constrained by public opposition. German policy on renewables has also been reinforced by concerns over global warming and environmental pollution. Following the UN Earth Summit Conference on Environment and Development in Rio in 1993, Germany committed itself to a 25 per cent reduction in carbon dioxide emissions.

The Commission of the European Communities

In addition to the support schemes provided by individual European governments, the Commission of the European Communities (CEC) also provides support, essentially backing up national programmes on renewables, as part of the European Union's response to the greenhouse effect and environmental protection. These programmes follow the basic stages of the innovation process, i.e. shifting from research and development and then through to commercial demonstration.

Thus research and development support is provided by the CEC's Joule scheme, which is part of (and has now been replaced by) the wider Fourth Framework R&D Support Programme, while its Thermie programme is aimed at providing support for commercial project demonstration schemes.

By 1989 the total allocated to renewables by the CEC had reached Ecu435.5 million (around £360 million) and it has continued to expand since then. Between 1990 and 1994 Joule was allocated Ecu262 million, while Thermie received Ecu700 million, although not all of this went to renewables. The new Thermie II programme for 1995–8 is expected to receive around Ecu200 million, while the Fourth Framework R&D Support Programme for 1995–8 includes Ecu1,000 million (around £800 million) for non-nuclear technologies, including renewables, via a new support scheme for Clean and Efficient Technologies (CEET).

In 1993 a new Altener support programme was added, with an Ecu40 million initial allocation for its first four years. The aim is to help create the necessary institutional networks and technical infrastructure (e.g. via information provision and training support) rather than to provide direct grants for technology projects. Overall it is hoped that it will help Europe achieve an 8 per cent energy contribution from renewables by 2005, with biofuels seen as playing a major role.

A 1993 CEC report 'The European Renewable Energy Study' (TERES) estimated that the European Union as a whole obtained only around 4.3 per cent of its energy from renewables. However, the report argued that if full account of environmental concerns was taken on board, by 2010 Western Europe could, on the most ambitious scenario, be obtaining up to 13 per cent of its primary energy from renewable sources. Note that this is the total *energy* contribution, not just the *electricity* element.

The pattern varies around Europe with Spain seen as having the potential to generate 20 per cent of its energy from renewables, Italy 23 per cent, but the UK only 9 per cent by 2010. The CEC report explains the relatively poor UK showing as follows:

> Although the UK boasts good potential for exploitation of renewables resources, it is also the EC's most self-sufficient fossil fuel user. Therefore, a long term strategy for energy is not seen as major government priority. With the demise of the Department of Energy in 1992 went the last remnants of such a strategy.

It adds

> The development of the country's full potential for renewable technologies can only be fostered within such a framework and thus is unlikely to be realised in the medium term, although the market imperfections introduced to favour renewables should allow the Government's modest aims to be met.
>
> (CEC 1993: 186 Annex 2)

Eastern and Central Europe

The CEC has become increasingly involved with initiatives in Eastern and Central Europe, and the TERES study also looked at the energy options there: it estimated that by 2010 Eastern and Central Europe might obtain up to 12 per cent of primary energy from renewable energy sources. The USSR had done some work on wind power and there were plans for major tidal energy schemes. However, while the energy resource is large, given the economic crisis following the break-up of the USSR, there are likely to be major development problems, not least in getting access to the requisite funding. Certainly, given the generally very polluting and inefficient energy generation systems that exist in many of these countries, support for renewable energy technology and improved energy efficiency ought to be a major priority.

Some Western aid has been forthcoming, much of it focused on cleaning up old fossil fuel and nuclear plants. In parallel there have been attempts to support the conversion of defence companies in the old Soviet bloc to non-defence related production, for example, via the CEC Konver Defence Conversion programme. In some cases these conversion projects have involved a shift to sustainable energy related projects. The Energy Committee of the European Commission has linked up with the UN Economic and Social Council to launch an Energy Efficiency 2000 project, designed to help stimulate the development and deployment of energy-efficient technology both in the West and in Eastern/Central Europe. As part of this programme a former military base at Ralsko in the Czech Republic has been converted into a diversification demonstration centre. Among other things, it is producing a range of energy-related devices, including heat meters for use with natural gas-fired district heating systems.

There have also been some private sector initiatives: in 1995 a former intercontinental ballistic missile plant in the Ukraine started producing US designed Kenetech 100 kW wind turbines of which 5,000 are to be installed in the Ukraine, followed hopefully by more elsewhere in Eastern and Central Europe.

Japan's renewable energy programme

Given Japan's role as a major player in the innovation field it is perhaps worth concluding our programme comparison with a brief review of how it has approached renewable energy development. Japan has few indigenous energy supplies and imports the bulk of its energy so it is hardly surprising that, following the oil crises of the mid-1970s, it has taken renewable energy seriously.

Solar heat collectors are now very common in Japan, as is the use of geothermal heat for hot water supplies and electricity generation. A wave power programme is also underway, with many small prototype oscillating water column units being tested, including some mounted on breakwaters. Although there are relatively few suitable sites for wind farms on the Japanese mainland, companies like Mitsubishi have been very successful in developing and exporting wind turbines. There are

Japanese equipped wind farms in California and Wales and Japan is opening a market for its wind systems around the Pacific rim area.

There has also been significant Japanese investment in photovoltaic (pv) solar cell development and a '70,000 roof' pv module deployment programme has been launched. Japan evidently sees pv as a major option for the future, possibly in conjunction with hydrogen production. For example, the electricity produced by pv modules could be used to generate hydrogen by the electrolysis of water and the hydrogen could be stored and subsequently used in fuel cells which, in turn, would provide electricity when needed. Some large (11 kW) fuel cell units are already in use. Up to 1991 some £62 million had been spent on pv R&D and a further £72 million has been allocated for the period up to 1995. Japan seems keen to explore the idea of the 'hydrogen economy', i.e. the production of hydrogen by electrolysis using renewable sources of electricity and its transmission to the point of use along gas mains, where it can either be burnt directly to provide heat or electricity or converted into electricity in a fuel cell. Hydrogen could thus become a new energy carrier, paralleling the transmission of power by electricity.

The development of renewables in Japan has been carried out under a variety of programmes, with government planning agencies working closely with private sector companies and universities. The 'Sunshine' alternative energy R&D project was started up in 1974 by the Ministry of International Trade and Industry (MITI) via its Agency of International Science and Technology. There was also a 'Futopia' wind power subproject: 'Fu' is the Japanese word for wind. A parallel energy conservation technology programme was set up in 1978, the 'Moonlight' project.

Overall MITI has played a key role in developing renewables, with collaborative private/public sector planning arrangements helping it to map out areas for strategic investment. A New Energy and Industrial Technology Development Organisation (NEDO) was set up in 1980 to 'promote the consistent and systematic transition from oil to alternative energy sources'.

In 1993 the Sunshine and Moonlight projects were integrated within a New Sunshine programme aimed at responding to the global warming problem by cutting Japan's carbon dioxide emissions by half by the year 2020, with alternative energy technologies and energy conservation expected to account for one-third of Japan's energy consumption. The New Sunshine programme's budget in 1994 was Yen52.8 billion and it is expected that by the year 2020 a cumulative total of some Yen1.55

trillion will have been allocated, equivalent to Yen55 billion p.a., or around £370 million p.a. (Bouda 1994).

Conclusion

The overall global renewable energy potential is considerable. A study produced for the UN Conference on Environment and Development in 1992 concluded that renewables could supply 60 per cent of the world's electricity and around 25 per cent of its heat requirements by 2025 (Johansson *et al.* 1993).

Even if very ambitious scenarios like these are discounted, the potential for renewable energy still looks very good and, as has been indicated, commercial, strategic and environmental considerations have led to increasing interest in renewable energy, along with energy conservation and energy efficiency, as a longer term option.

Chapter 8 reviews the overall state of play in the renewable energy field and then looks at some of the strategic issues and choices that may lie ahead if the world's large renewable energy resources are to be fully developed.

Summary points

- Renewable energy is being developed around the world, with the advanced countries taking a lead – notably the USA, Germany and Japan.
- The UK has a large renewable resource but is only developing it relatively slowly.
- Environmental concerns have been a major stimulant for these developments but equally there is growing awareness of the commercial potential of renewable energy systems.
- The potential of renewables looks very significant, but there is a range of strategic issues which will have to be addressed if this potential is to be developed fully.

Further reading

Inevitably, given the pace of technological change and policy shifts, the outline of renewable energy development given above will date. Useful detailed studies of global renewable energy potentials, including reports of some national level projects are include T. Johansson *et al.*, *Renewable Energy: Sources for Fuel and Electricity* (1993, Earthscan, London) and World Energy Council, *New Renewable Energy Resources* (1994, Kogan Page, London).

To keep up to date you could consult the annual *World Directory of Renewable Energy Suppliers and Services* (James and James, London) which in addition to listings of suppliers includes a series of overview papers reporting on the latest developments in technology and policy worldwide. Parts are available on the World Wide Web at http://www.jxj.com/dir/wdress/index.html.

The international renewable energy journal *CADDET*, produced in liaison with the International Energy Agency, provides useful reports of developments worldwide. It is available in the UK from ETSU at Harwell. It is also available on the World Wide Web at http://www.caddet.co.uk/.

Appendix II includes some other useful sources of information on renewable energy developments around the world.

8 Sustainable energy strategy

- Security of supply
- Compensating for intermittency
- Conservation versus renewables
- Scale and pace of deployment

A shift to a sustainable energy system would require a number of technical, economic and strategic issues to be addressed, not least the fact that some renewable energy sources are intermittent. This chapter looks at the current state of play on sustainable energy development and then at some of the emerging issues, including the problem of intermittency, the debate over whether to focus on energy conservation or on new energy supply technologies, and the question of the pace and timescale required for the development of sustainable energy systems.

The strategic state of play

Along with energy conservation and the more efficient use of the world's remaining fossil fuels, renewables look as if they may help the move towards a sustainable energy supply and demand system. Of course whether complete sustainability can be achieved in this way remains unclear: it depends on a range of issues, not least the level of economic growth that is being aimed at around the world. Wider issues such as these are explored in Part 3. But clearly the success or otherwise of any attempt to move toward sustainability will depend on the degree to which novel technologies in the renewable energy and energy conservation field can be developed and deployed. This chapter attempts briefly to summarise the current state of play in terms of sustainable energy development and then to look at some possible future lines of development. The emphasis is on **renewable energy**, but this is not to suggest that energy conservation is less important. There is no point in pushing ahead with new energy supply technologies unless energy demand is held down and, ideally, reduced.

Although, given the dominance of supply side thinking, energy conservation is still often underfunded, some progress has been made in recent years. This is in part because, aside from any environmental concerns, investment in the more efficient use of energy makes sense in economic terms. The technologies involved may not always be as exciting as the new renewable energy technologies, but they are important and for the foreseeable future this is where much of the practical day-to-day progress will be made.

In addition to the various technical means of energy saving, there is also a range of planning techniques which are likely to become increasingly relevant as part of a process of managing energy demand, such as least-cost planning and integrated resource planning. At the same time, much is happening on the supply side and later in this chapter we shall be discussing what might be the most appropriate balance between the supply side and the demand side.

Energy supply

On the supply side, as the previous chapter indicated, the trend around the world is clear: renewable energy technology is beginning to be seen as a major new energy option, with environmental considerations playing a major role in policy decisions.

For example, while Japan is still developing nuclear power it clearly sees renewables not only as a new export option but also as an important way to increase its energy self-sufficiency and reduce environmental impacts globally. In the USA, while nuclear power has in effect been abandoned, renewable energy systems are often seen as one of the cheapest new supply options, particularly when the comparative environmental costs of the alternative options are factored into the assessment, e.g. via the use of 'integrated resource planning' techniques. For many European countries environmental issues like global warming are often seen as particularly important, with renewables and conservation therefore being seen as increasingly relevant, especially since many have abandoned the nuclear option.

The result is that most countries are providing increasing levels of research and development support in key areas of renewable energy technology. Figure 8.1 shows the level of support for wind power in a range of countries. In addition many countries are providing support for

Figure 8.1 ***IEA governments' RD&D expenditure for wind energy 1983–91***

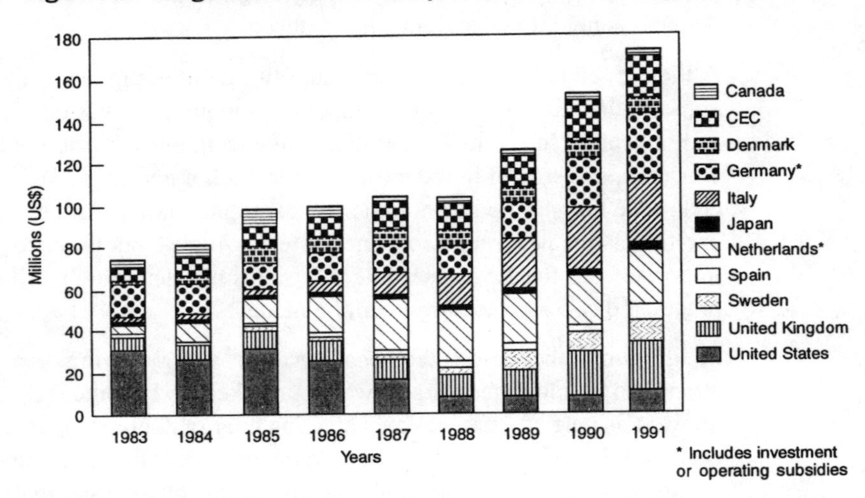

Source: World Energy Council, *New and Renewable Energy Resources*, Kogan Page, London, 1995, using IEA data.

commercialisation via some form of interim 'market enablement' scheme, whether through direct subsidies, grants or cross-subsidies like the Non-Fossil Fuel Obligation (NFFO). The main difference in these schemes, apart from relative size, is whether the funding comes via the state from taxes or from surcharges on consumers, this strategy being more common in countries like the UK where direct intervention and public support have been deemed politically undesirable.

The carbon tax

Underlying this general strategy is the assumption by funding agents such as governments that, given some initial market support, the technologies will mature and become fully commercially viable. At the same time it is assumed that the basic economic ground rules would change when and if the full social and environmental costs of conventional fuel usage are taken into account in the cost comparisons.

If concerns about global warming grow then it is possible that some form of economic penalty will be imposed by government legislation on the use of fossil fuels, reflecting their social and economic costs. For example the Commission of the European Communities (CEC) has proposed a 'carbon tax', i.e. a pollution tax on carbon dioxide emissions from fossil fuel combustion as a response to global warming. This would significantly improve the comparative economics of the renewable energy technologies – and also of nuclear power. One study suggested that if the full costs of energy were reflected in the price comparisons this would, in effect, bring forward by around ten years, the date by which photovoltaic solar cell technology reached commercial competitiveness and would dramatically improve the prospects of renewables generally (Hohmeyer 1992).

Developing sustainable systems

While the extent to which renewable energy technologies can achieve economic competiveness will inevitably be a key factor in shaping their progress in the short term, this is not the only issue likely to affect the development of renewables. Indeed short-term market concerns may be a poor guide if our aim is to move towards a sustainable energy system. After all, most of the benefits of renewables, particularly the larger scale ones like tidal power, are longer term. This is true in economic as well as environmental terms: it may take a decade or more before the initial outlay for a tidal barrage has been paid off, but the environmental and energy benefits and the income from its operation will last for centuries.

There is therefore a need to think about the development of the overall energy system in wider strategic terms. We have already discussed the idea that renewables should be given some credit for the role they can play in avoiding the emission of greenhouse gases. However, there are also other strategic issues. For example, a key element in any sustainable energy system must be that it is robust and can continue to provide power regardless of changed circumstances. Clearly, whatever the energy system, it is important to have secure energy supplies.

Security of supply

Security of supply can be ensured by having a diverse range of fuels and technologies and by selecting fuels which are not likely to be prone to interruptions in supply. Following the oil crisis in 1973–4, this was why many countries tried to diversify away from oil. There were also concerns about the impacts of industrial dispute on fuel supplies, following the various confrontations between workers and governments over the future of the UK coalmining industry.

More recently it has been argued that since renewable energy systems involve a range of scales of technology, using a diverse range of locally available indigenous sources, they are likely to offer a more reliable basis for secure energy supplies than systems which rely on imported fuels. Natural renewable energy flows are unlikely to be interrupted by human intervention, so there is no need to stockpile fuel. Terrorist attacks seem somewhat less credible with a national network of wind turbines than, say, with a handful of nuclear plants or offshore oil rigs.

Intermittency

As was noted earlier, some renewable sources are intermittent. The use of biofuels, hydroelectric power and geothermal energy can provide 'firm' (i.e. continuous) power. Tidal energy is very predictable but the availability of solar, wind and wave energy is not so reliable.

Fortunately some of the basic annual weather cycles link up well with human energy requirements. In much of the world it is windy during the winter, with wind and wave power at a peak when electricity is most needed for heating. Indeed there is a further correlation. Wind produces a chilling effect on buildings and conventional power systems have to take this into account by having extra generating capacity ready for windy days. In the UK this amounts to around 1 GW. However, if we have 1 GW of wind capacity on the system, its output will be correlated to the increased demand from the wind chill effect.

The shorter term variations in renewable energy input are more of a problem. The wind does not blow continuously so some wind turbines will be unproductive at any particular time. On average, as discussed in Box 3, Chapter 6, an individual wind turbine under UK conditions will only deliver near its full rated power for about 30 per cent of the time. Thus the actual generation capacity available from devices using these sources is less than the full theoretical capacity.

Perhaps surprisingly though, in practice this intermittency need not be a major operational problem if the electricity from these devices is fed into the national power grid network. So long as the total contribution from the various intermittent renewables does not exceed around 30 to 40 per cent of the total electricity on the grid, it can in effect even out local variations, so that the net overall power available from the grid remains more or less constant (Grubb 1991).

Up to this level, localised intermittency thus does not matter and the system can operate effectively without the need for energy storage systems. Beyond, say, a 40 per cent contribution there would be a need for backup supplies and/or some form storage, and this could limit the extent to which renewables could contribute. Short-term storage can be provided by a variety of mechanical and electric systems (eg compressed air, flywheel, batteries) but these are relatively expensive.

The hydrogen economy

However, one option currently being discussed is the idea of generating hydrogen by electrolysis, powered from renewable sources. Electrolysis provides a reasonably efficient way of using electricity to split water into hydrogen and oxygen, and the hydrogen gas could be stored or transmitted down conventional gas grids to where it was needed. In the interim, while natural gas is still available, it could be mixed with hydrogen. In the longer term there could be a shift in emphasis from electricity transmission to hydrogen transmission. Gas distribution is much more efficient than electricity transmission; gas can be stored more easily than electricity. When burned hydrogen produces only water as a byproduct.

A switchover to what has been called 'the hydrogen economy' could therefore have some attractions and there have been proposals for giant hydrogen gas grids around the world. Failing that, hydrogen could be tanked around in liquid form in ships: indeed this is already being done with hydrogen from Canada's hydro plants being shipped to Sweden. In the longer term countries in the Middle East might install large arrays of photovoltaic solar cells in desert areas so as to generate hydrogen to supply the less sunny areas of the world. Hydrogen can be used as a clean fuel for vehicles or can be converted into electricity in a fuel cell. It could thus be that in future there will be a global system based on renewable hydrogen, with electricity only being generated locally where and when needed.

Would this amount to just a very large-scale technical fix? That rather depends on how the system was designed, developed and run – and in whose interests. Technical possibilities like these seem likely to open up a whole new range of strategic debates concerning how systems should be developed in future.

Strategic choices

Perhaps the most urgent strategic issue is the question of the **pace** and **timescale** of renewable energy development.

Given that natural gas is cheap and reasonably clean, it might be thought to be premature and economically unrealistic to try to push ahead rapidly with the installation of renewable energy technology. It could be argued that, instead, the breathing space offered by gas should be used to research and develop the renewables more incrementally, prior to deployment on a wide scale.

There is certainly an environmental case for an incremental approach: it would give time for the social process of evaluating and negotiating trade-offs between local disbenefits and global benefits to be carried out in an open and participative way. At the same time there may also be an environmental case for more haste if, for example, the scale of the impact of climate change turns out to be significant. On this view it would be wise to press ahead quickly with renewables and move up the learning curve as fast as possible in terms of both technical development and social deployment. Of course this does not necessarily mean that a rapid 'crash' programme is needed: the vision should perhaps be radical and there should be a sense of urgency, but there are still merits in a careful, incremental approach even to the development of radical new technology.

Conservation versus renewables?

Another short to medium term strategic issue concerns the relative merits of renewables and energy conservation. Since, at least up to a certain point, energy conservation is usually cheaper and quicker than developing new generation capacity, what should be the investment priorities? The basic strategic criteria outlined earlier suggest that conservation should be preferred over generation, i.e. wherever possible investment should be targeted on techniques that cut demand, rather than on simply increasing supply.

However, as with some renewables, conservation initiatives can face implementation problems. While most of the initial energy efficiency measures may be relatively easy and cheap to deploy, some will require actions and purchases by individual consumers, and stimulating this type of initiative in a coherent and effective way is sometimes difficult. There is also a range of other implementation and uptake problems, not least institutional resistance to change, which may limit the extent to which the large, theoretically possible, energy savings can actually be achieved in practice in the short to medium term. We shall be looking at some of these problems in Part 3.

Hopefully these issues can gradually be resolved and conservation can play its vital role, along with renewables, in helping to move towards a sustainable energy supply and demand system. It is not a matter of choosing between renewables or efficiency: in general they are complementary. Both are required and both need to be deployed as rapidly as possible. However, there may be some specific tactical conflicts. For example, although in general conservation projects ought to be a prerequisite, in some circumstances it may be that renewable energy projects, especially smaller scale 'modular' projects, will prove to be more effective and easier to deploy. There may also be 'technical' conflicts: in a well-insulated building equipped with energy efficient end use devices, the level and pattern demand for energy is changed, so that the use of domestic scale solar power may be less cost effective. If there was widespread adoption of energy efficiency measures in all sectors then the overall pattern of demand would also change, and thus would influence the choice of renewables to some extent. Clearly what is needed is an optimal integrated approach and an appropriate mix between conservation and renewables.

There remain some more general strategic uncertainties as to the correct mix and right overall strategic emphasis. For example, it is sometimes argued that since the developed countries currently are the most wasteful, they should give energy efficiency a high priority, at least initially. Giving evidence to a UK government review on European energy policy, Greenpeace International proposed a 3:1 ratio, calling for a Europe-wide commitment to a 1 per cent p.a. increase in renewable energy use and a 3 per cent p.a. reduction in energy use via investment in energy efficiency (Greenpeace 1995).

An alternative view is that the developed countries should take the lead and give the development of renewable energy generation techniques the highest priority, both for their own use and, equally if not more importantly, by developing countries. Otherwise the newly industrialising countries will employ fossil fuel sources. On this view, since the advanced countries at present have the expertise, they should develop and then transfer renewable energy techniques for the developing world to use.

A hegemonic battle?

Clearly, the strategic perception influencing some of the views expounded above is that, regardless of the precise balance between them, renewable energy technology and conservation are *alternatives* to fossil fuels and that these must gradually, or perhaps even quite rapidly, displace and replace fossil fuels, as well as replacing nuclear power.

In strategic terms, the suggestion is that, in order to attain sustainability, it will be necessary to weaken the hegemony of conventional power supply technology, that is, its political, economic and institutional predominance must be challenged. Thus renewables must confront conventional supply options on their own ground, i.e. in terms of generation, while conservation must cut demand and open up crucial energy end use issues. Both can, to varying degrees, challenge the status quo of fossil and nuclear generation.

However, it is sometimes argued, most recently by Shell International, that there is no need for direct conflict between old and new technologies and that, in effect, there is room for all the various means of providing energy to expand, at least for a while, since overall global energy demand will increase in the decades ahead.

In Shell's 'sustained growth' global energy scenario (Figure 8.2), a summary of which was published in 1995, the use of fossil fuel, hydro and nuclear power are seen as continuing to expand until 2020 to 2030, at which point a plateau is reached. Subsequently, there are some reorderings within this pattern: oil and then coal begin to decline but gas stabilises, while nuclear and hydro still continue gradually to expand. At the same time, the 'new' renewables begin to lift off dramatically, supplying about 50 per cent of world energy by around 2060. However, overall energy demand rises by about a factor of three, so there is relatively little substitution of new renewables for the total fossil, hydro and nuclear fuel contribution, even by 2060 (Shell 1995).

Figure 8.2 ***Shell's sustained growth scenario (2% p.a.)***

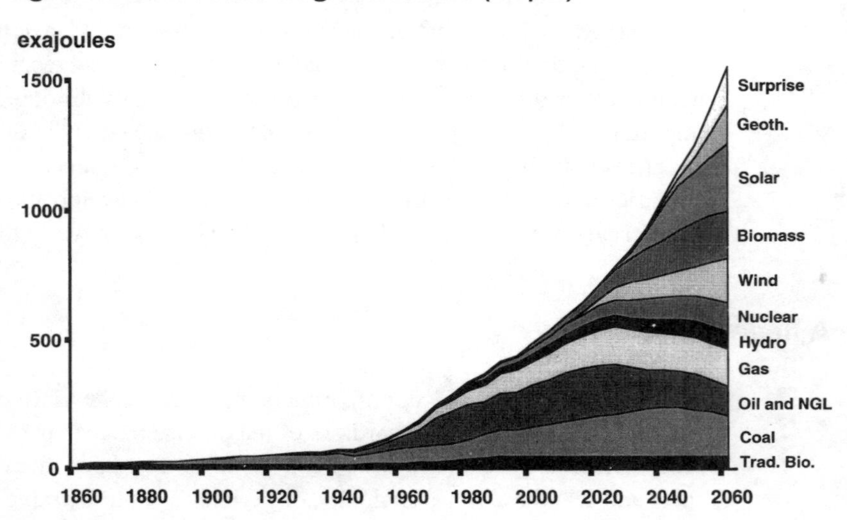

Source: Shell International.

In fact, far from substituting for the conventional energy sources, Shell seems to see the new renewables as something of an optional 'add on'. This becomes clear in their 'Dematerialisation' scenario (Figure 8.3) in which demand is further contained by a structural shift to a more energy-efficient world economy based on information technology and the use of less energy intensive materials. The result is that demand only doubles by 2060 and fossil and nuclear fuels supply the bulk of the energy needed. The lift-off point for the bulk of the renewables is postponed until 2040. Even then the growth of renewables is relatively slow and the 'second wave of renewables is not needed until 2060' (Shell 1995: 15).

Certainly, it will take time to develop renewables fully, so, in the interim, there will be a need for fossil fuel to avoid energy shortages and conservation to try to keep demand down and reduce the impact of conventional generation. But perhaps a more positive rendition of this strategy would be to portray conservation and fossil fuel technologies like natural gas CCGT and clean coal combustion as providing a bridge to the solar future, rather that seeing them as ways to avoid or postpone the need for renewables.

On this basis, and reflecting the need both for vision and a strategy for responding to global warming, we might tentatively add an extra and more aggressive strategic criterion to the list introduced in Chapter 3: *While continuing to invest in conservation and energy efficiency, get*

Figure 8.3 ***Shell's dematerialisation scenario***

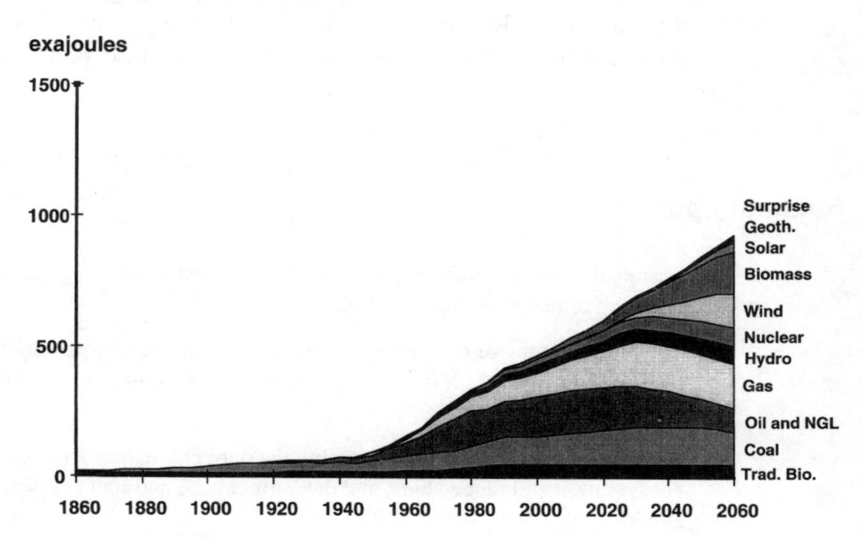

Source: Shell International.

on with developing sustainable energy generation technology as a matter of urgency.

This does not necessarily imply a crash programme of renewable energy development and certainly not one at the expense of conservation. But it does imply that renewable energy R&D should be significantly expanded and rates of deployment increased as far as is practicable, in order to ensure that a transition to sustainabilty is not delayed by vested interests in the existing energy system. Clearly this is a matter of interpretation and political judgement, but it is a view that is becoming increasingly influential among lobbyists and in political circles (Scheer 1994).

As has been indicated, the level of commitment to renewable energy development varies around the world, although it seems overall to be growing. Thus, although it may be felt that an effort should be made to promote the use of nuclear energy, one of the key conclusions reached by the usually relatively conservative World Energy Council, at the World Energy Congress in Tokyo in 1995, was that the level of support for the development of renewable energy should be increased. It added:

> The next two or three decades represent the key period of opportunity for transition to a more sustainable path. But unless we begin the process now, the various development paths for the long term are likely to get blocked off.
>
> (World Energy Council 1995)

For the immediate future the key requirement would seem to be to get moving on the development and deployment of the new sustainable technologies and to start the long process of learning how to respond to the various implementation problems which will inevitably emerge. It is to these problems that we now turn – in the next part of this book.

Summary points

- Sustainable energy technologies may increase security of supply, e.g. by reducing reliance on imported fuel.
- The intermittency of renewable energy sources can be compensated for by the use of integrated grids and also possibly by switching to a hydrogen economy.
- There are strategic debates concerning the correct balance between conservation and renewables, and the correct pace and scale of deployment generally.

- There may be a need to move quickly to develop and deploy sustainable energy systems, but there is likely to be a range of implementation problems.

Further reading

A relatively conservative but still very positive view of the strategic requirements for renewable energy development is taken in the World Energy Council's *New Renewable Energy Resources* (1994, Kogan Page, London). The Shell International study mentioned in the text, *The Evolution of the World's Energy System 1860–2060* was produced in 1995 (Shell, London). A more radical viewpoint is adopted by the President of the Eurosolar lobbying group, Hermann Scheer, who is a member of the German parliament, in his forceful *A Solar Manifesto* (1994, James & James, London).

For general updates on current debates, policy issues and technological progress, particularly in relation to the UK, see *RENEW*, the journal produced by NATTA.

Part 3 Problems of implementation

Sustainable energy technologies are being developed and introduced worldwide. However, there can be obstacles to this process. Part 3 looks at some of the technical, economic, strategic and environmental limitations and problems facing the development and implementation of sustainable energy options. It focuses in particular on the institutional difficulties of obtaining support for the development of renewable energy technology, and on the problems of winning acceptance from the public for the use of such novel technologies. Although renewable energy technologies are relatively benign in environmental terms compared with conventional energy technology, they may still have local environmental impacts and adverse public reactions can be an obstacle to implementation.

9 Getting started: institutional obstacles

- **Research and development problems**
- **Institutional resistance to change**
- **The need for 'follow through'**

The development of new technologies is never easy. This chapter looks at the difficulties experienced by renewable energy technologies in trying to get funding for research, using the UK wave power programme as an example. It also looks at the way in which wind power has been developed around the world, highlighting the different research and development strategies that were adopted. Finally it looks at the technical problems experienced by the UK geothermal energy programme, highlighting how prone novel projects can be to initial failures.

Obstacles to sustainable energy development

New technologies face a whole range of technical, economic and institutional hurdles as they try to get started on the long process of development and deployment. These constraints have been particularly apparent in the case of sustainable energy technology and they may well shape the extent to which these new energy technologies can play a role in supporting a move towards a sustainable energy future – or at least they may influence the pace at which this transition occurs.

The focus in this chapter is on the early stage of the research and development process and, as a consequence, the emphasis is on renewables since, in the main, this is where most R&D effort is needed. Energy conservation techniques are relatively well established, the main problem is to get support for deployment. We shall be looking at deployment and implementation problems facing both conservation and renewables in the next chapter.

Here we are more concerned with the initial getting started stage and in particular with the problems which researchers have experienced in obtaining initial government financial support for research on renewables. As will be seen, this has not always been easy and even when forthcoming the resultant projects may not always fare well and support may not be continued.

Getting started

Most new technologies need initial R&D support to get established, and obtaining access to this represents a major institutional hurdle, especially since the potential source of initial support is likely to be from the government. Inevitably there are disagreements about which projects should be funded and over how projects should be developed. Certainly it is hard to 'pick winners' when technologies are at an early stage of development. In addition there may be resistance to new developments from those with vested interests in the existing range of technologies and a lack of commitment from decision-makers to pressing ahead with what may seem risky and long-term development programmes. In effect we are facing the reality of the sort of battle for hegemony that was mentioned in Chapter 8.

Rather than trying to analyse the institutional problems facing renewables in the abstract, it may be helpful to focus on some specific examples. We now look, via a series of mini case studies, at the problems which have faced a range of renewable energy projects. We start with a brief review of the way wave power was treated in the UK, this being one of the more celebrated, or perhaps notorious, examples of lack of follow through.

The UK wave energy programme

As noted in Chapter 7, in 1974 the Labour government launched a programme of renewable energy research and development. Although not generous this has continued under successive UK governments, allowing a range of technologies to be developed and/or assessed.

Wave power was initially seen as a front runner. Glyn England, then Chairman of the Central Electricity Board, suggested in 1978 that it might offer enough power 'to supply the whole of Britain with electricity at the present rate of consumption' (England 1978: 272).

When a Conservative government came to power in 1979 it evidently remained supportive. John Moore, then Energy Minister, commented in September 1980 that 'whatever other problems wave energy researchers may face, lack of Government support will not be among them' (Moore 1980). The device teams in universities and elsewhere worked enthusiastically; some scale prototypes were tested in open water.

However, by 1982 views had changed and wave R&D was cut, following assessment of all renewables (apart from tidal, which was examined separately) by the government's Advisory Committee on Research and Development (ACORD). The government had asked ACORD to assess the renewable options as part of its regular review process, in the light of expected cutbacks in overall renewable energy R&D funding (from £14 million to £11–12 million) and budgetary expenditure.

David Mellor, then Secretary of State for Energy, later commented that it had not proved to 'be possible to insulate this area from the savings the Government are making in public expenditure' (Mellor 1982).

The justification for cutting back on wave power was economic. The Energy Technology Support Unit, which reviewed wave power and the other renewables for ACORD, had put the likely cost of wave power generation at 4–12 p/kWh and had concluded that 'wave power is likely to be economic only in those futures more favourable to renewable energy technology' (ETSU 1982: 14).

The full story of the ACORD review and subsequent events is complex and opens up some interesting general issues, for example, concerning the way technological innovation is handled (Elliott 1995). But the basic strategic issues are relatively clear. The approach adopted in the UK government's initial programme on wave power, from around 1976 onwards, was to establish a 2 GW 'reference design' target and ask the wave power device teams to develop proposals. Given that only tank-tested models were available at this stage, it was obviously quite a jump to try to come up with designs for full scale systems which might have 20,000 tonnes displacement or more. Subsequently there was some criticism of this approach as premature – asking too much too soon of an embryo technology. One commentator likened it to 'designing a super tanker whilst at the same time developing the principles of naval architecture' (Flood 1991: 44).

Small is beautiful?

It has been argued by some innovation theorists that large-scale projects like this are inevitably inflexible and do not allow for piecemeal adaptation, incremental development, feedback and learning from mistakes. Thus, in a 1993 study of the UK wave power programme, Audley Genus argued that the 2 GW wave system design target reflected the UK energy establishment's obsession with large-scale units, which he saw as having 'all the hallmarks of inflexible technology'. Thus, in the case of the wave programme, 'lead times appear to be long, capital intensity high, unit size large' and 'enormous investments of time, capital and other resources would have to have been made before any learning about actual performance and improvements of these systems could be realised' (Genus 1993: 141).

This in turn made it very difficult to come up with sensible cost estimates and even harder to make sensible decisions about the future of the technology. Stephen Salter, the inventor of the 'duck' wave power system, commented that it was like trying 'to decide our aviation policy on the data available in 1910' (Salter 1981a).

Beyond that there have been allegations that errors were made in ACORD's assessment and that some of the more favourable assessments by outside consultants were suppressed. A fierce debate ensued on the way wave power had been treated. Some critics suggested that there had been a pro-nuclear bias and concerns were expressed about the location of ETSU at the Atomic Energy Authority's Research establishment at Harwell. For its part, ETSU strenuously protested its independence. Other critics claimed that, at the very least, the technology had been assessed at too early a stage in its development. As the all-party House of Commons Select Committee on Energy put it in 1984, the suspicion was that wave energy 'was effectively withdrawn before the race began' (Select Committee 1984: xxxii).

Certainly there was room for disagreement among the experts; they faced the problem of trying to cost a range of very novel systems. As a spokesman for the Department of Energy was to tell a subsequent session of the Energy Select Committee, there was 'definitely scope for different judgements at the early stage of the development of a device' (Select Committee 1992: 125).

There was considerable pressure from the Select Committee and from renewable energy lobbyists for a reassessment and in 1989 a new review by ETSU was set in motion. However, when the results eventually emerged in November 1992 the conclusions were similar:

deep sea wave energy was still not seen as economic and inshore/onshore wave energy systems were not viewed much more favourably (Thorpe 1992).

Of course it could be argued that, since very little new work had been done in the ten years since the 1982 decision to cut R&D support, this conclusion might not be surprising. Nevertheless, the result was that wave power remained in the 'long shot category' as far as the UK government was concerned and this has continued to be its subsequent position. Although there have been some signs of industrial interest in smaller scale in/onshore wave power, the 1982 decision certainly put a halt to work on deep sea wave power.

Developing technology

The wave power story seems to indicate that new technologies face major problems in getting accepted: there can be institutional biases and a lack of vision. Thus, commenting on the way in which the wave energy programme had been assessed in 1982, the House of Commons Select Committee on Energy noted that there was no evidence that the Department of Energy had 'ever assessed the much larger expenditures on the fast breeder reactor and fusion research against similarly stringent cost criterion – certainly not at such an early stage in their development' (Select Committee 1984: xxxii).

Of course, equally, it could be said that in the case of wave power the negative assessments were right in the end, although it is hard to prove this either way at present. In effect the jury is still out. Given the uncertainties, and regardless of the rights and wrongs of the way wave power was treated in the UK, it is obviously difficult to decide how to develop new technologies. Much of the basic data are just not available and there are few precedents, especially with novel technologies like renewables; those that do exist may be unhelpful.

These problems are well illustrated by the second mini case study concerning the way in which wind power was developed, focusing mainly on the approaches adopted in the USA, UK and Denmark. In this case our emphasis moves from the initial research phase to the development phase and then on to commercial deployment.

The development of wind power

From the mid-1970s the USA adopted a 'high tech' aerospace approach to wind power development, with the emphasis on increasingly large complex prototypes like the 2.5 MW Boeing/NASA 'Mod 2' series. Germany and Sweden also embarked on similar projects. When the UK started up its own wind programme in the early 1980s it adopted a similar approach. After building a smaller 250 kW prototype, a large 3 MW machine was installed on the Orkney islands off the north of Scotland by the Wind Energy Group. This eventually cost around £17 million.

However, the large machines were not successful: there were some technical failures due, for example, to the great stresses on the giant blades. In general, after the initial enthusiasm they were seen as too big, complex and expensive, pushing the technology too far, too soon. Around the world commercial emphasis shifted to smaller machines of around 300 to 400 kW rated power.

Denmark had already taken a lead by initially emphasising relatively simple, robust, machines based on proven agricultural engineering approaches. Local agricultural engineers, operating almost on a craft basis but with carefully targeted state support, developed small wind turbines which subsequently proved to be world beaters. The Danish machines were sold in great numbers to the USA and later to the UK and have been gradually scaled up as technical experience was gained and markets extended. Subsequently Japan, the USA and UK have developed similar designs.

The argument against starting with large-scale projects in relation to wave power thus clearly applies also to wind power. Most of the large complex wind turbines initially developed in the USA (by NASA, Boeing, etc.), in the UK (by the Wind Energy Group consortium which included British Aerospace) and elsewhere (e.g. in Germany where a 6 MW unit was constructed) have now been abandoned.

Part of the problem with the 'top down', large-scale 'high tech' approach was that aerospace engineering concepts were not as relevant as expected when it came to devising systems that had to operate with dramatically varying wind loads. As one US wind turbine engineer put it: 'We were guilty of steady flow aerospace thinking, and largely did not appreciate the range and difficulty of the wind environment' (Stoddard 1986).

The eventual outcome illustrates the weakness of the research-led programme – adopting the so-called 'technology push' approach as opposed to being geared to respond to 'market pull'. Of course, in the early years there were no markets for wind turbines but as they emerged 'technology push' gave way to 'demand pull'. Denmark was fortunate to be in a position to exploit this, for example, on the basis of the rapidly expanding US market and for a while enjoyed a 90 per cent share. Subsequently Japan entered the world market with devices of similar design.

Interestingly, machine sizes are now increasing, with 500 to 700 kW machines in use and some 1 MW units being developed. Incremental development seems to be the order of the day. Certainly, the approach to initial machine development adopted in Denmark was far less expensive than that adopted in the USA – the Danish wind industry received a total of $52 million via government grants during the initial phase, while the equivalent figure for the USA was $450 million (Karnoe 1990).

Overall then it seems that the initial focus on small machines, developed on the basis of an incremental cut and try 'bottom up' approach, has clearly triumphed over the 'top down', high tech approach and has laid the basis for expansion.

Technical success and failure

The wind turbine example shows how technological development patterns interact with market economic mechanisms and clearly economics is a central issue for renewable energy technologies. One way in which the institutional and economic problems can be reduced is if technological progress can be made which may, for example, help to reduce generation costs and build more confidence in the technology. Technological breakthroughs are possible, as are incremental improvements. Equally, however, there can be technological problems which may dissuade funding sources from continuing to back a new technology. The risks may be perceived as too great.

This seems to have been the case in our next example – concerning the fate of geothermal 'hot dry rock' technology in the UK.

The UK geothermal programme

The potential geothermal resource is thought to be quite large, representing, if fully developed, perhaps 10 per cent of UK electricity requirements. The UK's hot dry rock programme involved an experimental well at Camborne in Cornwall, funded by the government. A double well system was created, but the initial results were disappointing: less power was produced than expected due to problems with the geology. Establishing an efficient heat-extracting well configuration is a difficult and expensive art, even assuming the basic concepts are sound.

As you may recall, the idea is to create a fissure system between the bottom ends of the two wells to act as means of collecting heat, with cold water being pumped down one well and hot water/steam emerging up the other. The fissure system is usually created by exploding a small charge at the bottom of the wells. However, this has to be done in exactly the right way.

To put it simply, if the fissure system that is created allows water to pass through too easily, the flow is too high and not enough heat is picked up. If the artificial heat exchanger created by the fissure offers too much resistance to the water being pumped though, the flow rate is reduced and again not enough heat is absorbed. The fissure connections have to be just right and this depends on the precise geological nature of the strata. One might therefore expect failures before a successful well was achieved, much as oil exploration companies accept that many early drillings will prove fruitless.

Nevertheless, the initial failure of the Camborne well led to a loss of confidence in the project, which by this stage had cost some £42 million. In 1994 the UK programme was halted, with the government concluding that the technology was not likely to generate electricity economically. Work on geothermal power has continued elsewhere, but the Camborne project has been portrayed as a failure (National Audit Office 1994).

Conclusion

Technical problems like those experienced by the Camborne geothermal project are not uncommon at the early stage of the research and development process: initial technical problems are to be expected as part of the learning process. The withdrawal of financial support may have as much to do with short-term economic considerations as actual technological problems. It could be argued that a longer term perspective is required: the large geothermal resource around the world is unlikely to be ignored for long.

Fortunately, renewable energy technology has usually been able to move up the learning curve, as illustrated by the wind power case study, and as in the case of photovoltaics and many other renewable energy technologies.

Figure 9.1 ***Cost of electricity from wind in the USA***

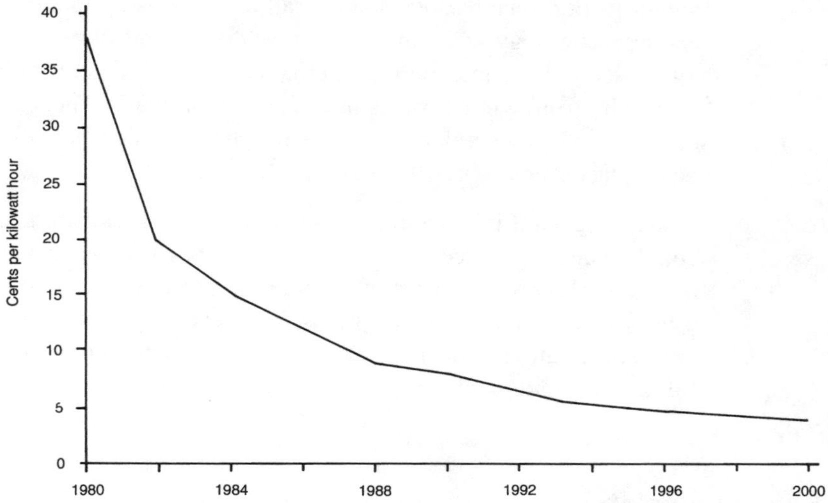

Cents per kilowatt hour
40
35
30
25
20
15
10
5
0
1980
1984
1988
1992
1996
2000

Source: Unpublished data from the US Embassy. Reproduced in the Select Committee on Energy Report on 'Renewables Energy', Session 1991–92 Fourth Report, Vol.1: xiii, xiv, HMSO, London, 1992. Crown Copyright, reproduced by permission of the Controller of HMSO.

Figure 9.2 ***Cost of electricity from photovoltaics in the USA***

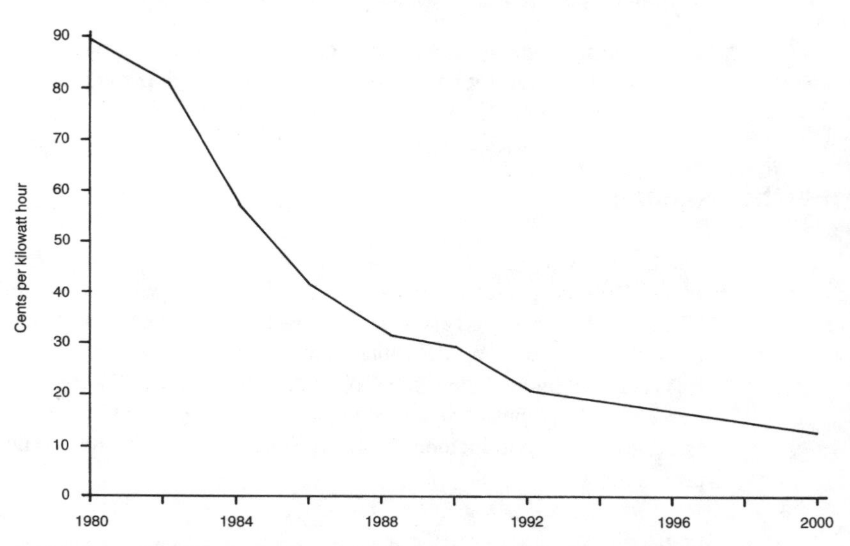

Cents per kilowatt hour
90
80
70
60
50
40
30
20
10
0
1980
1984
1988
1992
1996
2000

Source: Unpublished data from the US Embassy. Reproduced in the Select Committee on Energy Report on 'Renewables Energy', Session 1991–92 Fourth Report, Vol.1: xiii, xiv, HMSO, London, 1992. Crown Copyright, reproduced by permission of the Controller of HMSO.

Some projects, like deep sea wave power, may be stalled for economic and/or institutional reasons, but overall it seems likely that renewable energy technology will continue to develop in terms of performance and reliability and become more cost effective. For example, the costs of electricity from wind turbines have fallen by around 70 per cent within a decade or so, and similar reductions for photovoltaic solar cells have taken place (see Figures 9.1 and 9.2).

However, even if it is possible to obtain institutional support for research and then develop successful and economically attractive technologies, there are many other problems facing renewables in trying to become established as major energy options. The next chapter looks at the problems relating to the next stage in development – getting support for full-scale deployment.

Summary points

- Obtaining support for research and innovation is often difficult.
- Vested interests may resist the development of new technological options.
- Bottom-up development strategies may be more effective than top-down approaches.
- There may be lack of support for 'follow through' when projects are faced with initial problems.
- Even if new energy technologies can pass successfully through the research phase, there can still be problems with full-scale implementation.

Further reading

The UK wave power story is analysed in detail by David Ross in *Power from the Waves* (1995, Oxford University Press, Oxford). The wind power story in the UK, USA, Denmark and elsewhere is covered by Paul Gipe in *Wind Energy Comes of Age* (1995, Wiley, Chichester). The UK renewable energy programme was reviewed by the independent National Audit Office, *The Renewable Energy Research, Development and Demonstration Programme* (1994, HMSO, London). This report includes useful critiques of the wind and geothermal programmes.

10 Keeping going: deployment problems

- **The problems of 'short termism'**
- **The need for support infrastructure**
- **The need for public acceptance**

Even when novel energy technologies have passed through the research phase successfully, it is sometimes hard for them to get support for full-scale implementation. The existing financial and institutional arrangements and priorities often do not favour new technologies. This chapter looks at some of the financial problems that have faced renewables, using tidal power as an example. The difficulty of promoting energy conservation is also examined, using the experience of the UK's Energy Saving Trust as an example. Finally, the question is introduced of how public acceptance for novel energy technologies can be gained – since this is the ultimate hurdle in the implementation process.

The problems of transition

The momentum behind the development of sustainable energy technology seems fairly well established. For example, the economic benefits of energy conservation are clear and it seems likely that renewable energy technology will increasingly come to the fore as the environmental costs of existing energy technology become more apparent. The global renewable resource potential is large and the economics of the conversion technologies are beginning to look reasonable: for example, wind power is now seen as competitive with conventional sources in some contexts and costs are continually falling.

On the assumption that a commitment is made to a sustainable energy future, Figure 10.1 presents one view of how the global energy use

Figure 10.1 ***Renewable energy scenario: percentage of total global primary energy demand***

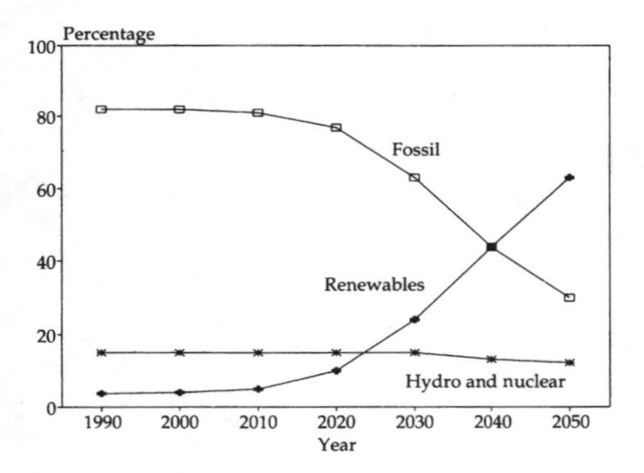

Source: D. Anderson, *The Energy Industry and Global Warming*, Overseas Development Institute, 1992.

pattern might change in decades ahead. A start in this direction has already been made.

However, the pattern of development so far has revealed a number of major institutional problems which may make a rapid transition to a sustainable energy future difficult. These are not always trivial problems: they involve a battle over the whole paradigm of technological development. In 1981, before his funding was cut, Stephen Salter, the wave energy pioneer, commented:

> We are attempting to change a status quo which is buttressed by prodigious investment of money and power and professional reputations. For 100 years it has been easy to burn and pollute. 100 years of tradition cannot be swept away without a struggle. The nearer renewable energy technology gets to success, the harder that struggle becomes.
>
> (Salter 1981b: 580)

Chapter 8 looked at some of the key elements of various possible strategies for sustainable energy development, and it was clear that there could be conflicts over the direction, balance and pace of development. Chapter 9 looked at some specific examples of these types of problems: it is hard for new technologies to get started, for example, in terms of obtaining funding for research.

Unfortunately, even assuming that the necessary level of financial support for research and development is obtained, and that the technology can be developed successfully, there still remains a series of problems relating to the next stage – the practical full-scale deployment of renewable energy systems and energy efficiency programmes. These difficulties are the concern of this chapter.

Problems of deployment

A key problem facing new technologies and techniques is that they are inevitably trying to establish themselves in an institutional, market and industrial context based on the existing types of energy technology. There are powerful vested interests in the status quo and in many cases this is reflected in the current financial, organisational and institutional environment, which may not be very well suited to the acceptance of novel developments. There is, arguably, something of a mismatch between the new technology and the existing support infrastructure, and a disinclination therefore to provide the necessary funding.

To some extent this may be surprising. After all energy conservation has obvious economic advantages and, superficially at least, renewable energy is a free resource – with most natural energy flows there are no fuel costs. So they ought to be commercially attractive. However, all change involves disruption and the economic advantages may take longer to materialise than financial backers are willing to accept. As illustrated in Chapter 4, the 'payback time' problem is often less in the case of energy conservation measures, but it can be serious for some renewable energy technologies. Although the fuel is essentially free, at least in the case of natural flow sources like wind or solar energy, there are nevertheless significant costs associated with constructing the necessary energy conversion technology.

With conventional power plants, although construction is expensive, a significant part of the overall cost is in the fuel, and this cost is incurred after the plant has been built. So some of the cost is deferred until after the plant has started earning its keep. By contrast, with renewables since the fuel is free, at least with natural energy flow based plants, there is only the upfront capital cost of constructing the system. So, rather perversely, it is harder to find investment sources.

Nowhere is this more clear than in the case of **tidal power**. Therefore it is perhaps worth looking briefly at the fate of tidal power in the UK which, as has been noted, has some of the world's largest tidal energy resources. Tidal barrages are not a new concept: the technology has been proved and a barrage has been operating successfully in France, on the Rance estuary, since 1968. However, obtaining finance for major tidal projects proved to be difficult in the UK.

Tidal barrages in the UK

Initially, when the UK renewable energy research and assessment programme started in the mid-1970s, the tidal option was seen as quite significant. The UK had some of the best sites in the world, notably the Severn estuary. A government-backed Severn Barrage Committee was set up to review the potential and it reported back in 1981, confirming the results of several earlier studies which had indicated that a tidal barrage on the Severn estuary would be technically feasible and could generate power at competitive costs, once built.

Although the Severn barrage scheme was seen as technically viable, the capital cost of building it would be very large. The Severn Tidal Power Group, an industrial consortium consisting of many of the UK's leading construction and engineering companies, developed a proposal for an eleven-mile long barrage which would have 8,640 megawatts of installed generating capacity and would be able to meet on average about 6 per cent of UK electricity requirements. But the capital cost would be around £10 billion. Government supported feasibility studies nevertheless went ahead, culminating in the Severn barrage project report (Department of Energy 1989).

At this stage much of the emphasis was on the likely environmental impact of the Severn barrage scheme. Some environmental organisations had objected to it since they felt it could adversely affect wildlife and the local ecosystem generally, although, as the scheme's promoters pointed out, some of the social impacts could also be positive, e.g. the provision of a new Severn road or rail crossing, increased opportunities for watersports and local employment opportunities.

As it turned out, the main problem to face tidal power was not the potential local impacts on the natural environment but the changed economic environment which emerged following the privatisation of the UK electricity industry in 1989–90. The Central Electricity Generating Board, the UK's nationalised utility, was broken up and the smaller private companies that replaced it were unlikely to be interested in a major project of this sort.

A project on this scale might have been viable as a long-term, publicly financed national investment, or perhaps via some partnership arrangement with industry, since public sector rates of return would then have been all that would be needed. However, following the privatisation of the electricity industry, it became clear that the government was not going to providing any further funding and the Severn barrage project was stalled. The private sector would be unlikely to want to take it on singlehanded since, in the economic climate of the time, it would be looking for much higher rates of return over shorter periods.

Attempts were made to investigate smaller, less expensive tidal barrage options, notably on the Mersey. The Mersey Barrage Company, another consortium of major UK companies, tried to obtain NFFO support for their scheme. In the end however this attempt was unsuccessful and the Mersey project and several other smaller barrage proposals were abandoned. In 1994 the tidal programme was more or less wound up, after around £14 million had been spent (Department of Trade and Industry 1994b).

As can be seen, the primary problem in the case of tidal barrages was not technological: the technology existed and was relatively mature. The problem facing the UK barrages was finance, and the relatively short-term economic perspective that prevailed in the energy sector, as elsewhere, following the privatisation of the UK electricity sector. Once built, barrages would pay off their investment costs in a matter of decades, after which, like hydroelectric plants, they would produce very cheap electricity for centuries. However, the initial capital cost is large and since the technology was relatively mature, significant cost reductions were unlikely. As a result, investment funds proved impossible to find: the payback times were simply too long.

Supporting innovation

Large projects like tidal barrages clearly have problems in obtaining funding. But although smaller scale renewable energy products may have the attraction that they can be deployed incrementally on a modular basis, similar problems can also face them. The difficulty is essentially one of institutional preconceptions, for example, as to what represent suitable areas of technological investment by financial institutions. In the past the emphasis in terms of major financial investment has been on the use of large-scale concentrated forms of energy, managed by large-scale centralised agencies. Investment agencies are therefore often suspicious of smaller projects, which are sometimes viewed as likely to be risky, low yield investments.

This has been particularly true in the UK where some wind projects have found it difficult to get finance, given the high rates of return expected by private sector investors, and the risks they feel that these novel technologies involve. Certainly some of the pioneers in the renewable energy field have found it hard going. For example, the first wind farm in the UK was established by Peter Edwards, a Cornish farmer and landowner. The project was given support under the government's non-fossil fuel subsidy scheme, but this only provided an income when power was produced: it did not give 'upfront' money for construction. He had the benefit of a 40 per cent EC capital grant, but this was conditional on finding the rest of the capital elsewhere which he achieved, in part, by selling off his dairy herd.

In some other countries the technology and infrastructure are better matched and this may provide an indication of how progress can be made in future. For example, wind power has been very successful when developed and deployed on a small local scale, as in Denmark. The locally produced Danish wind turbines sold in large numbers around the world. Perhaps equally important is the fact that this export success was launched from a strong domestic market which had been created from the bottom up. As noted earlier, around 70 per cent of the 3,000 wind turbines installed in Denmark so far are locally owned by co-operative guilds. Such local ownership and development has meant that funding was relatively easily obtained; local banks were willing to provide loans.

Interestingly, another major renewable energy source using fast-growing energy crops like willow or poplar also looks as if it can best be developed at the relatively small scale local level. This renewable energy source seems likely to fit in well with existing rural agricultural, economic and social patterns and practices. For example, banks are accustomed to providing loans to farmers and the technical infrastructure for growing and harvesting crops already exists on farms.

This discussion highlights a general point which has emerged from current theorising on innovation: the successful deployment of new technology requires the existence or the development of suitable social and institutional contexts – a technical infrastructure, suitable financial networks, a skill base. These may not yet exist for all of the renewables, but seem to be emerging for some of the smaller scale, modular and locally scaled options.

Supporting energy conservation

The same points also apply in the case of energy conservation, possibly with even more force. As has been noted, there can be problems with deploying energy efficiency measures, for example, due to the fact that they often require responses by many independent consumers. In this situation it is vital to have support networks, providing advice and, where necessary, offering grant aid to stimulate uptake.

There have been many attempts to establish such networks and support schemes. There is also a range of institutional mechanisms, like integrated resource planning, which can be used to try to ensure that when investment decisions are being made by companies the relative merits of investing in energy saving and energy supply are given equal consideration. This approach has sometimes proved particularly fruitful in countries like the USA, where some power companies have found that they can obtain better financial returns by selling their consumers energy efficiency packages than by investing in new energy supply technology.

However, in many countries energy conservation initiatives still tend to be given low priority. This has been particularly so in the UK. Following the privatisation of the energy supply industry, the institutional structure of the supply side meant that it was in the interests of the various generating and distribution companies to sell more of their product, in competition with each other. Attempts have been made to try to resolve this problem – so far without success. The fate of one of these UK initiatives, the Energy Saving Trust, provides an example of how institutional conflicts can bedevil sustainable energy projects.

The Energy Saving Trust

The Energy Saving Trust (EST) was established by the Conservative government in 1992 and was seen as part of its response to the commitment made at the 1992 UN Earth Summit to stabilising UK carbon dioxide emission by the year 2000. An overall UK savings target of a 10 million tonne of carbon (mtC) was agreed and the Energy Saving Trust was expected to be able to deliver savings of around 2.5 mtC by the year 2000. To do this it would support a range of energy efficiency projects – with funds raised by a levy on consumers' bills, thus avoiding the necessity of having to raise a direct tax. The sums required would be large, rising to £150–200 million p.a. and totalling perhaps £1.5 billion by the year 2000. But by 1993 all seemed set for a range of projects, including the provision of subsidies to domestic consumers to offset the cost of buying very efficient gas-condensing boilers and low energy light bulbs, a series of local combined heat and power projects and the creation of a network of local energy advice centres.

At this point disaster struck. As part of the privatisation process, the government had established a series of regulatory bodies charged primarily with stimulating competition in the various sectors of the energy market, in the belief that this would drive prices down. The gas and electricity regulators (OFGAS and OFFER) saw the proposed EST levy as essentially an indirect form of taxation and the proposals were blocked. OFGAS and OFFER argued that they had been set up to keep prices down; if the government wanted to support a programme of energy conservation in pursuance of its environmental policies, it should legislate for direct taxation.

There followed much institutional wrangling. A limited amount of money was allowed through for the EST to use but by 1995 the government had found that, in fact, the necessary carbon savings were likely to be made in other ways – notably as a result of the so-called 'dash for gas' in the electricity generation field. As was noted earlier, following privatisation, combined cycle gas turbines were increasingly installed as an alternative to the coal-fired plants, with a resultant significant reduction in carbon dioxide emissions. The large EST programme was therefore not needed.

The EST operation was drastically scaled down: rather than 2.5 mtC the target fell to 0.3 mtC, with only around £29 million p.a. being available as opposed to the original plan for a £150–200 million p.a. programme (Owen 1995).

As can be seen it proved difficult to develop an innovative institutional approach in a context defined by powerful vested interests in conventional approaches. The government's ideological commitment to competition had effectively collided with its commitment to carbon saving, but the latter was achieved, in the end, by the acceptance of expansion of supply based on using gas.

Energy conservation was therefore the victim: the government clearly felt that it was really up to consumers to be alert to any savings they could make, and that the market should not be distorted by subsidies.
As for resolving environmental problems, it could be claimed that the 'dash for gas' provided an example of what the market could do if left alone. The EST continues to operate, albeit at a much reduced level, and the local energy advice centres have survived, but the longer term future for this programme is unclear.

Social acceptance

The existence of advice centres and technical and financial support networks is of course only part of what is needed for successful deployment of sustainable energy technologies. For new technologies to prosper there is also a need for the technologies to be accepted by the general public.

Social acceptance is particularly important in the case of energy conservation initiatives in the domestic sector, since their success depends on the adoption of new technologies and techniques by consumers. Information and advice centres can alert people to what is available: but uptake has often proved to be a problem: there is still evidently a need to convince consumers of the benefits.

At first sight social acceptance would seem less of a problems in the case of renewable energy technology. Opinion surveys indicate widespread popular support for the development of renewable energy technology in general terms. For example, 87 per cent of the respondents to a Gallup poll in the UK, carried out in 1991 for Friends of the Earth, indicated that they would prefer the government to increase spending on renewables. Even so, when it comes to specific projects there may be objections, with adverse local environmental impacts being a key issue.

However this depends on the context, and in particular the economic context. As the next chapter will illustrate, there have been some strong objections to wind farms in the UK, where the machines have been introduced, as it were, by absentee owners. But so far there has been very little public opposition to the siting of the locally owned wind farms in the Danish countryside: as the Danes say 'your own pigs don't smell'. The wind projects promoted by the Danish utility companies have, by contrast, sometimes met with opposition.

In principle, compared to most conventional energy generation technology, renewable energy systems like wind turbines ought to have some advantages in terms of public acceptability. They are relatively small scale and although there may be a need for fairly large numbers of them, in contrast to the hidden dangers of, say, a nuclear plant, they have the advantage that 'what you see is what you get' – they are functionally transparent. Their purpose and operation is usually clear from their appearance and there are no hidden or longer term environmental problems. For example, if necessary wind farms can be easily decommissioned and removed, thus returning the site to its original state.

However, the widespread introduction of renewable energy technology like wind turbines could involve significant changes in land use and landscapes and might even involve changes in social patterns, for example, a degree of decentralisation. So if renewables are to develop on a significant scale then it is vital that they gain public support.

Put less passively, it seems vital that the public can influence the way they are developed and deployed. The early enthusiasts for alternative technology argued that one of its attractions was that it could be developed and used on the smaller scale, and might therefore be more susceptible to local level democratic control. Not all renewable energy systems fit this description, but some do and it will be interesting to see what role local control can play in reality in shaping the way renewable energy develops.

This process may not always be easy or without conflict, for example, in terms of local environmental impacts. These conflicts provide an interesting display of the problems of 'thinking globally and acting locally'. Certainly, the development of some renewable energy technologies is likely to be constrained by local environmental planning and land use factors.

Equally, the deployment of renewables may be stimulated by increasing environmental concerns over the generally much more significant global impacts of using conventional energy technologies, e.g. global warming from the emission of greenhouse gases like carbon dioxide produced when fossil fuels are burnt. The local and global impacts have to be traded off against each other.

Environmental conflicts

Economic concerns inevitably determine the success or otherwise of technological projects, but land use issues and environmental concerns also enter into the commercial equation – if for no other reason than because it takes time and money to obtain planning permission. So while conventional economic factors are obviously important, the need for a trade-off between local and global environmental factors is also important and may shape the way in which renewables are developed.

Renewable energy provides part of a solution to global environmental problems, but no technology can be entirely benign in environmental terms. Although the impacts of most renewables are relatively small and localised compared to the often large and global impacts of conventional energy technologies, there can still be local problems. Perhaps the most significant impacts are associated with large tidal barrages and hydro plants, which involve large civil engineering constructions and large modifications to local and even regional or national ecosystems and natural energy flows.

Studies of public attitudes to proposals for a barrage on the Severn estuary in the UK have indicated some mixed responses (Barac *et al.* 1983). There was a general enthusiasm for renewable energy projects and support for the local economic benefits which the advent of large projects like this could bring, not least local employment. Equally however there was concern over local impacts on wildlife and the ecosystem generally, combined with more instrumental concerns about any adverse effects on the ease of navigation by shipping.

Although wind farms involve less environmental modification, they also intrude on the landscape. The first wind farms, installed in California in the late 1970s and early 1980s, were generally well received but there were also some objections. Many of these related not so much to the visual appearance as to concerns that the machines were not really economically viable. There were suspicions that they were being installed without much concern for their technical success, simply to take advantage of the tax credit system being operated by the Californian state government as an incentive for wind energy development. Certainly many of the early wind projects turned out to be technically unviable and the planning controls on their siting were, in some situations, fairly lax. In essence, what had happened was a 'wind rush', somewhat like the 'gold rush' a century or so earlier. This rather chaotic phase nevertheless had some benefits: it provided a context in which new designs could be tested and certainly the technology developed rapidly. So too did awareness of the need to win public acceptance. In some cases this required no more than the provision of better information. But it also became clear that there was a need for careful siting and layout to minimise visual disruption and for sensitive local consultation (Thayer and Hanson 1988; Gipe 1995).

As the case study in the next chapter illustrates, this became even more clear when the first wind farms were built in the UK, where population densities are generally much higher. The problem of noise pollution also became important. Technological developments are helping to reduce some of the problems. For example, the new generation of variable speed wind turbines is less noisy. But it will still be important for developers and planners to be sensitive to local concerns and to consult with communities over the location and layout of proposed schemes.

To summarise, renewable energy technologies present system designers and planners with an interesting challenge: they must try to balance the global advantages of renewables (in terms of, for example, reduced emissions of greenhouse gases like carbon dioxide) against the local impacts, and come up with technically workable, economically viable and environmentally acceptable compromises. What seems to be needed is some way to negotiate a balance between global and local.

The wind farm case study

The next chapter presents a detailed case study on public reactions to wind power which in effect represents the 'first shots' in this process. The aim of the case study is to provide an insight into how the political process of conflict and negotiation works or might work in practice, as discussed in Chapter 1. The choice of wind power for this case study should not be taken to imply that wind farms face any major problem of public acceptance either in the UK or elsewhere. As will be seen, even in the relatively crowded UK the majority of wind farm proposals have been successful and, overall, wind power has proved popular. However, the wind farm debate does provide a useful illustration of the problems facing land-using renewable energy technologies.

Case studies could have been chosen from many other areas, for example, based on local concerns around the world over emissions from waste combustion plants; the conflicts that have emerged in some locations over the development of geothermal energy systems; or the debates on ecosystem impacts in relation to large-scale hydroelectric projects and tidal barrages. But the debate over wind farms has the merits of being well documented, particularly in the UK, and involves a smaller scale technology whose nature and likely impact is relatively easy to understand.

Summary points

- Some large renewable energy projects face an uphill battle in terms of obtaining finance for full scale deployment since the benefits are sometimes longer term.
- Smaller scale projects may be able to obtain local finance and be more publicly acceptable.
- There is a need for public awareness of the relative costs and benefits of sustainable energy technologies.
- Public concerns have to be addressed and local reactions have to be considered in the process of developing sustainable energy technologies.

Further reading

There has been a long running debate over the impact of renewable energy, particularly in relation to wind farms and tidal power. The pros and cons of tidal power are discussed by Clive Baker in *Tidal Power* (1991, Peter Peregrinus/IEE, London). See also Carol Barac, Liz Spencer and Dave Elliott, 'Public awareness of renewable energy: a pilot study', *International Journal of Ambient Energy*, 4 (4):199–211, 1983, which looks at reactions to the proposed Severn barrage.

The UK wind farm debate is the subject of Chapter 11. A useful starting-point for a discussion of reactions to wind farms in the USA is provided by R. Thayer and C. Freeman, 'Altamont: public perceptions of a wind energy landscape', *Landscape and Urban Planning*, 14: 379–88, 1987. The classic study of impacts is Alexi Clarke's report 'Wind farm location and environmental impact' (1988, NATTA, Milton Keynes).

There are some useful overview papers from a UK perspective in the January 1995 issue of *Land Use Policy*, 12 (1) including 'Energy, land use and renewables' and 'Renewable energy and the public' by Gordon Walker.

You can keep up to date on the ongoing strategic and environmental issues relating to the development of renewables in the UK and elsewhere by subscribing to the bimonthly journal *RENEW*, produced by NATTA (see Appendix II).

11 Case study: public reactions to UK wind farms

- **Conflicts between developers and local communities**
- **The role of planning**
- **The need for sensitive consultation and siting**

New energy technologies are inevitably unfamiliar and their deployment can lead to public concern – particularly if they are perceived to have negative local impacts. This chapter presents a case study on the often heated debate over the location of wind farms in the UK. It provides an example of the need, when seeking to introduce new projects, to be sensitive to local concerns and at the same time to try to balance local environmental costs against global environmental benefits.

Introduction

The UK wind farm programme, which got underway from 1990 onwards, owes its existence primarily to the non-fossil fuel cross-subsidy scheme introduced following the privatisation of the electricity supply industry.

As was noted in Chapter 7, a Non-Fossil Fuel Obligation (NFFO) was imposed on the newly privatised regional electricity supply companies (RECs), requiring them to buy in set amounts of electricity from nuclear and, to a much lesser extent, renewable suppliers. In addition, a surcharge was imposed on fossil fuel electricity generation in order to meet the extra cost of using non-fossil sources, with this cost being passed on by the RECs to their customers.

Two NFFO orders were set specifically for renewables in 1990 and 1991, with in all 197 projects being offered a 'premium' price over and above the usual 'pool' price for their electricity. The structure and constraints of these NFFO orders played a major role in shaping the initial pattern of wind farm development and, arguably, public reactions to it. The central problem was that the NFFO cross-subsidy scheme was seen as likely to be an infringement of the EC's rules on fair competition. Some EC members were also opposed to providing support for nuclear power. Consequently, as a compromise, a 1998 deadline was imposed for the NFFO scheme, with renewables, in effect, being inadvertently penalised. As we will see, subsequently, for NFFO 3 and thereafter this constraint was removed for renewables. However, the 1998 deadline played a significant role in shaping the initial development of wind power in the UK: it meant that intending developers could only receive the premium cross-subsidy price for a limited period which was reduced as the cut-off date approached (Elliott 1992).

The result was something of a 'wind rush', with perhaps insufficient time allowed for full environmental assessment. The 1998 cut-off date also meant that the premium price offered had to be quite generous and artificially enhanced. Wind projects were offered 6p/kWh in the 1990 NFFO and 11p in the 1991 NFFO, in order to allow companies to recoup their investment in the shorter period remaining. This had the effect of making wind power look very expensive and raised the spectre of 'profiteering'. Wind farm developers were sometimes seen as rushing in to exploit 'handouts' without paying sufficient attention to environmental impacts.

There may be some truth in this; certainly developers with good windy sites may have benefited from the fact that all the projects received the same basic average payment regardless of position. Equally, however, it could be argued that developers had little choice but to try to find windy, upland sites. Even given the relatively high NFFO levy payments, for most developers the 1998 cut-off date meant that it was hard to get financial backing. Their profit margins were tight and most evidently felt compelled to target the high wind speed sites. These would have been financially tempting in any case, but the 1998 deadline provided a further impetus. As we shall see, given that these upland sites were sometimes in environmentally sensitive areas, this led to significant local objections.

Reactions to wind farms

As the practical deployment of wind projects picked up speed from 1991–2 onwards, data on public reactions became available. Before and after responses to the first project – a 10 Vesta turbine scheme at Delabole in Cornwall – were carried out by consultants the government's Energy Technology Support Unit (ETSU). The 'before' study was carried out in 1990 and the 'after' study in mid-1992, six months after start-up. A 'control' study of opinion in Exeter was also carried out.

The conclusion was that basically wind power was popular. Only about a third offered not-in-my-back-yard (NIMBY) – responses, and support for wind farms consolidated when people had local experience. For example, around 25 per cent of those who were initially concerned about the Delabole scheme changed their minds: 80 per cent said it had made no difference to their day-to-day lives; 44 per cent approved; and 40 per cent approved strongly. On visual intrusion, more than 40 per cent had thought it might be a problem in 1990, whereas this had fallen to 29 per cent by 1992. On noise, only 14 per cent had not expected this to be a problem in 1990, but by 1992 around 80 per cent felt it was actually not a problem.

Overall, the introduction of the wind farm had 'altered attitudes in the direction of local residents being more favourable towards wind energy', with 'many of the worries local residents had about wind turbines' having been proved unfounded (ETSU 1993b: 54). Interestingly, Delabole had around 100,000 visitors in its first year.

Subsequent studies have shown similar patterns of support. For example, the wind farm developers National Wind Power, reported strong local support for its wind farm at Cemmaes in mid-Wales. Of local people asked 98 per cent 'liked or didn't mind' it while, according to Ecogen, another development company, a survey of 500 local adults carried out by a local school near the Coal Clough wind farm found that 70 per cent wanted more wind projects. Interestingly, it also found that although 40 per cent of the people sampled could not see the wind farm from their homes, 55 per cent of them would have actually preferred to be able to see it. Similarly positive reactions emerged from a survey of 1,000 residents by a local Friends of the Earth group in Sidlesham on the Manhood Peninsula: 83 per cent supported the wind farm proposal there; 7 per cent were against; and 10 per cent were unsure.

However, wind projects are clearly not popular with everyone. There were some bitter planning battles as the UK programme got underway with, as we shall see, local opposition to some projects becoming very significant.

The planners and conservation groups' responses

The first wave of wind farm projects faced local planners with considerable problems. There were few precedents for this type of development, e.g. were they to be seen as agricultural projects, as reflected in the label 'wind farm', or were they industrial projects, i.e. power plants? Consequently there were requests for new planning directives from the government. Draft (consultative) Planning Guidelines emerged from the Department of the Environment in 1991. However, in the main they left the detailed assessment up to the local planning authority; the basic policy being to support wind for global strategic reasons unless local costs outweighed them. In effect the planners were asked to balance the national and global benefits against any local disbenefits.

However, local council planners were in general unhappy with the lack, as they saw it, of clear guidance. Thus, responding to the Draft Planning Guidelines, the Association of District Councils suggested that without 'clear statements of strategic need, the Local Planning Authorities may lack a sufficiently strong case to justify the inclusion of positive proposals for wind farms . . . in local plans, against the probable weight of local objections'. They pointed out that 'Local planning authorities are not responsible for energy generation' (Association of District Councils 1991).

The County Planning Officers Society added that 'it is doubtful if individual local planning authorities can realistically be expected to consider the overall environmental benefits when the actual rewards remain intangible' (County Planning Officers Society 1991).

Some conservation and environmental groups went further. The Countryside Commission and the Council for the Protection of Rural England felt that the Planning Guidelines indicated that the government was being 'soft' on developers. By contrast, some of the more radical pressure groups saw the problem as being not so much lack of control of and/or profiteering by the developers but a failure by the government properly to support wind power. Thus the Campaign for the Protection of Rural Wales indicated that, while they would oppose any specific project they felt was ill-conceived, they would also 'lobby hard to amend the financial package currently on offer to developers', so as to allow for less invasive siting (Evans 1991).

For its part, the government tried to avoid making a link between planning and financial issues. Thus, in a letter to Friends of the Earth (28/3/91) Colin Moynihan, then an energy minister, denied any link, claiming that the siting issue was 'fundamentally a planning issue rather than a commercial one'. The full Planning Policy Guidelines (PPG22) emerged in February 1993, but they differed only in detail from the draft and did not really resolve the issue. Responding to continued pressure for clarification, in November 1993. Tim Eggar, then an energy minister, spelt out the government's position as follows:

> The Government does not have a specific target for wind energy, and its success will depend on developers finding sites acceptable to the public and to planning authorities. NFFO does not override the planning process and Government is as concerned as much about the local environment as the global one. The planning guidance in PPG22 requires planners to balance the Government's policies for renewables with those of the countryside.
>
> (Eggar 1993)

Making this type of judgement is obviously difficult and, in principle, each case was to be judged on merit without setting a precedent. Nevertheless, the first few planning inquiries did seem to follow a common pattern.

Public inquiries

The first Public Inquiry, in 1991, was on the Wind Energy Group's 24-turbine wind farm proposed for Cemmaes in the Dyfi Valley in mid-Wales, on the edge of Snowdonia. Despite objections by the Countryside Commission, among others, it led to very positive recommendations from the inspector, whose conclusion was subsequently accepted by the then Secretary of State for Wales, David Hunt. Although the visual impact issue was seen as relevant, Hunt felt that it was 'not sufficiently compelling to outweigh the need for renewable energy' (Hunt 1991).

The next inquiry focused on WEG's proposed 15-turbine wind farm on Kirkby Moor in Cumbria, just south of the Lake District National Park. The inspector turned it down on the grounds of visual intrusion, but his conclusions were overruled by the Secretary of State for the Environment who argued, 'such harm as may be caused by the visual impact of the wind farm in this instance is outweighed by the national need for the development of alternative cleaner sources of energy' (Department of the Environment 1992).

Most of the applications were not called in for full planning inquiries and, at least initially, the majority obtained planning permission, even if in some cases this had required the intervention of the Secretary of State. For example, by the end of 1993 John Gummer, the Environment Secretary, had overturned four rejections by local planners. However, this was not a fixed pattern. After strong local opposition, the Welsh Office turned down an appeal relating to an application for a National Wind

Power farm on Anglesey; and more than fifteen other proposals have been turned down by local planners. Even so, the objectors and those environmental groups which were opposed to the wind farms evidently felt that they were fighting an uphill battle.

Thus, a local solicitor who had acted for objectors to Ecogen's proposed wind farm on St Breock Down in Cornwall commented, 'everyone believed it would be unthinkable in the face of opposition from just about every quarter. The inspector just steam rollered over the objectors' (Key 1994).

The environmental groups' response

The response from the environmental groups was mixed. Friends of the Earth, locally and nationally, maintained their long-held support for and promotion of wind power. So did the Labour Party affiliated Socialist Environment and Resources Association (SERA), and Greenpeace. However, several of the other major conservation groups came out against wind farms, while the Campaign for the Protection of Rural Wales, which initially adopted a supportative if critical stance, subsequently changed sides. So did the Ramblers Association, while the Welsh Tourist Board's 1994 'Tourism 2000' report expressed concern over the impact on tourism.

The Countryside Council for Wales (CCW) came out with a particularly strong opposition line arguing that while wind turbines 'are welcome as a source of renewable energy, the scale of their contribution to meeting energy needs does not justify overturning established planning policies and safeguards' with wind power projects tending 'to threaten precisely those areas that CCW is charged to protect'. They also added that there 'should be a presumption against wind turbine development in areas of close proximity to sites with the benefit of statutory landscape designation status' (CCW 1992).

The English equivalent, the Council for the Protection of Rural England (CPRE), also expressed critical views. For example, in the CPRE's evidence to the Department of Energy's Renewable Energy Advisory Group, it called for greater scrutiny of projects and suggested that wind power should not be seen as a technical fix for 'the key political, social and economic problem – the profligate use of energy' (CPRE 1991).

Subsequently, Tony Burton from the CPRE told the *Guardian* (11/3/94) that while they were not opposed to wind power in principle, 'the system of subsidies is putting pressure to build wind farms in quite inappropriate places, remote landscapes that have been protected for decades'.

To summarise then, while some environmental groups supported wind power, others moved to oppose it, and some evidently felt that, in the absence of proper government policies, they were having to 'police' the developers' projects unaided.

For their part some local planners resented being asked in effect to make national energy policy related decisions without, as they saw it, proper guidance. In response, some have argued that it is beyond their competence and remit and they therefore focus only on local issues. This has resulted in the rejection of some wind farm proposals, which subsequently may have been overturned. This is not to say that local planners or all the conservation groups were necessarily against wind farms in principle. The objections were to the pace of development and lack of guidance.

The mood of some planners was well expressed by Tim Horwood, planning officer for Cornwall County Council, which produced its own interim policy guidelines: 'Instead of so many schemes opening at once it would have been better if they could have evolved more slowly so there would have been more time to consider all the implications' (Horwood 1992).

The wind backlash

Given that relatively large numbers of project proposals were coming forward, it is perhaps not surprising that some conservation groups moved into opposition. Some were clearly concerned that the programme was moving too fast. Thus the Northern Devon Group of the CPRE saw it as a 'mad scramble' for lucrative hilltop sites, with wind turbines threatening 'to stride across the countryside of North and West Devon like a plague of triffids' (Allen 1991). Barry Long from the Countryside Council for Wales told *The Times* (21/8/91), 'It will be hard to stand on a hilltop in mid Wales without seeing a windmill'.

During 1993 objections emerged around the country, e.g. in Devon, Cornwall and Yorkshire. But perhaps the largest number emerged in Wales. The UK's largest wind farm so far, at Llandinam, proved to be something of a turning-point in the debate. There had been objections to its scale, on visual intrusion grounds, but in the event it was noise that proved to be the major problem. Several local residents claimed to be suffering major disturbance, and there does indeed seem to be a significant noise problem for some residents in the valley below the ridge on which the 103 Mitsubishi machines are sited (Walker 1993).

The local experience with this project stimulated significant objections to subsequent projects in Wales, for example, in relation to National Wind Power's (eventually successful) application for permission to install a 22-turbine wind farm at Bryntitli. This helped to consolidate local opposition generally. A Noise Action Group was formed locally and subsequently a National Windpower Consultative Association has been set up, initially covering the Welsh border areas.

An extensive debate ensued in the local press, with opinions often becoming polarised. The local responses have occasionally thrown up some bitter invective. One unsigned leaflet, circulated in mid-Wales in 1993, talked of Wales being 'in the forefront of being covered in swathes of ugly turbines to line the pockets of foreigners and greedy owners' (NATTA 1994a), although, in general, the debate has been carried out in less aggressive terms.

Even so, some colourful allusions have emerged. The Ecogen wind farm at Llandinam was claimed by a resident to sound like a 'twin-tub' washing machine. National Wind Power's wind farm at Llangwyryfon was alleged to sound like 'an old wheelbarrow being pushed along continuously', while their Cold Northcott project in Cornwall was described as sounding like 'a huge washing machine gathering speed to spin dry' (NATTA 1993a).

Clearly, noise was the major issue for these people, which is a difficult matter to address. Visitors are usually surprised at how quiet wind farms sound, just a slight blade swish, even close up, together with occasional gear train rumble. However, some machines are noisier than others and in some topographical situations these sounds can evidently be amplified by resonance effects, e.g. by valleys. People's responses to the result can also vary. Some are very sensitive to low grade background noise (and cannot sleep with a fridge running). Certainly, once a noise starts to be annoying it can be detected, even at very low levels. Some individual machines may have been particularly noisy during their run-in periods. The developers have been trying to respond to noise problems generally, e.g. by installing noise insulation materials in the housing of the machines.

Less can be done about visual intrusion once the wind farm is established. So far this seems to have been less of a problem: once the machines were in place most people seem to have got used to them. However, clearly not everyone has accepted the idea. 'Lavatory brushes in the sky' was how Sir Bernard Ingham described the wind farm near Hebden Bridge in Yorkshire, an allusion which has subsequently been repeated in various forms by the media. As one-time press secretary to Margaret (now Lady) Thatcher and public relations adviser to British Nuclear Fuels, Ingham has extensive media contacts and has been very outspoken on the wind farm issue.

With the involvement of such major public figures, the wind farm issue began to take on a national perspective and gained considerable national media coverage. Perhaps the key event was the setting up early in 1993 of a national anti-wind lobby group, Country Guardian, dedicated to opposing 'the desecration of the coasts and hills by wind farms', with Sir Bernard Ingham as vice president. At the start of the campaign Joseph Lythgoe, who set up Country Guardian, sent out 1,800 letters, 'one to every weekly newspaper in the country'. Since then the group has been very successful in gaining media attention.

Subsequently a campaign against the proposal to site 44 turbines at Flaight Hill, near Hebden Bridge, has also attracted national media attention and the involvement of a number of celebrities, including pop star Cliff Richard and many notable literary figures. They wrote a letter to the *Times Literary Supplement*, with 62 signatories, complaining about what they saw as an 'assault on our literary and artistic heritage', given that the wind farm would be in Brontë country.

The media's response

The press has generally showed considerable interest in the local debate over wind farms. All the major national newspapers (*The Times*, *Guardian*, *Independent*, *Telegraph*, *Observer*, *FT*) have carried reports, with coverage increasing as objections mounted and the emphasis mainly on the negative side.

The broadcast media have generally also adopted a fairly critical and, in some cases, hostile approach. The BBC's 'Country File' television programme (24/4/93) included a fairly critical review, while the Radio 4 'You and Yours' programme (30/7/93) presented a more or less unremittingly negative view. A Radio 4 'File on Four' programme (8/3/94) was a little more hopeful, although it did suggest that the development of wind farms might turn out to be harder than developers and environmentalists had initially hoped.

The local press in the relevant areas carried regular news stories and features plus extensive letters. A rough survey of coverage in mid-Wales during the period between March and December 1993, although in no way exhaustive, may be indicative of the general pattern: there were sixty-two news items, nearly all reporting 'problems', and thirty-eight letters of which only eight were pro-wind (NATTA 1994b).

Most of the objectors cited specific local problems – noise and visual intrusion, but some reflected wider conservation and preservation concerns, as well as fears about the impact on tourism. Some of the supporters complained about the 'lack of balance' in the media debate, with Ian Mays from the British Wind Energy Association complaining to the *Guardian* (5/11/93) that 'a small but vociferous number of people have generated a disproportionate amount of press coverage'. Nevertheless, the campaigns clearly had some effect. According to the Guardian (9/3/94):

> Tim Eggar, the Energy Minister, is believed to be alarmed at the number of objections to proposed wind farms by groups who claim that their turbines impose noise and visual blight on the landscape. The DTI has been inundated with protests as part of a campaign at local and national level orchestrated primarily by . . . the Country Guardians.

A DTI spokesman had earlier denied that any fundamental shift in policy was likely, telling the *Financial Times* (22/12/93) that 'we don't need to radically change our policy because of opposition, because we have always said that the go-ahead for developments is dependent on the relevant planning permissions'.

Even so, it seemed likely that wind projects were going to be much more carefully scrutinised in the next round. NFFO 3 had been heavily over-subscribed, with more than 650 renewable energy project proposals, including many wind projects. On 11 March 1994, just after the deadline for final applications had passed, Tim Eggar commented:

> Although some 230 of these proposals are for wind energy I would expect to see no more than twenty or so wind farms result from the next round of the NFFO. But this will depend on developers' abilities to find sufficient windy sites which are acceptable in planning terms, particularly from the point of view of noise and visual impact. If they fail to do so, or to get nominated projects up and running, this will count against wind energy in future rounds.

Reflecting the media campaign, he added:

> I can understand the concern expressed by some about the number (of) wind farms submitted for the latest round of contracts. I am prepared to see the steady development of wind energy, but it is not the case, as suggested by some alarmists, that I intend to cover the country with wind farms. The holding of a

> NFFO contract does not bestow any special consideration. Government planning policy requires the environmental benefits of wind and other renewables to be taken into account, but also places just as much emphasis on the need to protect the local environment.
>
> (Eggar 1994)

Validity of the objections

Mike Harper, then director of the British Wind Energy Association, asserted that 'the controversy to date has largely revolved around misconceptions and misinformation distributed by groups aiming to stifle wind energy development completely' (Harper 1994).

Certainly there have been cases of misrepresentation and even disinformation: for example, the 10 per cent fossil fuel levy has sometimes been cited by wind farm objectors as being the extra cost imposed on electricity consumers by the wind farms. In fact the bulk of this 10 per cent is due to the support provided for nuclear power. Even given the initial artificially high level of support that had to be provided to wind projects as a consequence of the 1998 NFFO deadline, the wind farm element so far has still only added around one-tenth of a per cent extra on an average consumer's bills.

While it seems unreasonable to suggest, as did one national columnist, that 'conventional rural opinion is now probably, if anything, anti-wind power' (Ridley 1994), the opposition cannot be written off as simply being mistaken or as just reflecting effective lobbying by a small minority of activists. Even though the opinion polls seem to indicate generally high levels of support (see Box 5) there is also clearly genuine local opposition in some areas. The lobbyists could not operate effectively unless there were objections at the grass roots level, and the extensive local press coverage does indicate that there is no shortage of local people who are seriously concerned.

Unfortunately, however, the effectiveness of the anti-wind lobby in obtaining press coverage, the polarisation of views and often bitter tone of the ensuing debate have made it difficult to judge the actual scale of opposition. This said, even though it will be a biased sample, press cuttings do give some idea of the pattern of views. Simple NIMBY-type responses would seem to predominate, but there are also sometimes wider regional preservation and conservation concerns.

Some objectors, more subtly, accept the global argument (e.g. in relation to greenhouse emissions) but claim that wind farms cannot help that

Box 5

Results of some public opinion surveys on wind farms

Kirkby Moor

A study of 250 local residents near the 12-turbine wind farm at Kirkby Moor in Cumbria was commissioned in February 1994, six months after start-up by National Wind Power. It revealed that:

- 82 per cent supported the development of wind farms in the area and 84 per cent thought that more energy should be generated from renewable sources.
- 83 per cent were 'not at all concerned' or 'not very concerned' about the noise that they make.
- Of those who could see the wind farm from their houses, 77 per cent were 'not at all concerned' or 'not very concerned' about the impact on the landscape.

Taff Ely

A study of 250 local residents near the 20-turbine wind farm at Taff Ely in Wales was commissioned in February 1994, six months after start-up, by East Midlands Electricity. It revealed that:

- Only 2 per cent strongly opposed the development of wind farms in the areas.
- 75 per cent said that either they could not think of any disadvantages of wind power or there were no disadvantages.
- Noise was not perceived as a problem, with only 3 per cent saying they could hear the wind farms from their homes.

Source: Data relayed by the British Wind Energy Association 1994

much, so that the local impact is not justified. Others favour alternative energy options – energy conservation often being seen as a better choice. For others the central issue is what they see as 'profiteering' by 'greedy developers' who make use of the 'extensive subsidies' without, allegedly, being concerned with the impact on the local environment.

On the local proponents' side, strong support for wind farms is seen as part of a positive commitment to the future, with the threat of 'global warming' often being cited, along with the dangers of nuclear power. Many supporters also say they actually like the look of the wind farms. On the other hand, those who do not obviously feel strongly about it. Thus *Newsweek* (28/3/94) quoted Sir Bernard Ingham as saying 'people who think they're attractive are aesthetically dead'.

Clearly views on aesthetics differ, so too do views on environmental intrusion. Some of the responses reviewed in this case study are of the NIMBY kind, i.e. not-in-my-back-yard parochial protectionism. Other responses are more general. There has been a growth of environmental concern over the last few years in response, in part, to the campaigns of environmental movements on issues such as global warming and acid rain. However, this new level of sensitivity may be giving rise to some contradictions, in the form of local level resistance to one of the remedies proposed for the global problems, i.e. wind farms. Thinking globally and acting locally, as the slogan goes, may not always be as easy as expected.

Summary points

- The economic constraints in the early NFFO rounds led to some invasive siting and precipitated some negative local reactions.
- Familiarity generally leads to more acceptance, and overall support was relatively high, but pressure groups can create an impression of widespread objection.
- Sensitive consultation and careful choice of sites is vital if public acceptance is to be obtained.

Further reading

The three-volume report of the Welsh Affairs Select Committee on 'Wind Energy' (Session 1993–4 Second Report , HMSO, London, July 1994) gives a good overview of the debate on wind farms in the UK. Volume 1 summarises their conclusions.

The Department of Environment's Planning Policy Guidelines (PPG22) can be obtained from HMSO, London. ETSU can also supply various guides for planners and developers.

12 Public acceptance: the need for negotiation

- **Local opposition to renewable projects**
- **The need to negotiate public acceptance**
- **Developing 'social control' of technology**

The impacts of renewables are generally much less than those of conventional energy technologies, but there is still a need to negotiate public acceptance. This chapter reviews the wind farm case study and looks at how the debate has continued, with the focus moving on to attempts to ensure that local communities can have more direct involvement with, and benefits from, such projects. Local involvement is vital in that, rather than seeing local concerns as a problem, well-informed criticism might also be seen more positively as an attempt to subject technology to some form of direct social control. After all, one of the alleged benefits of at least some types of renewable energy technology was that they were likely to be more amenable to local democratic control than the preceding large-scale, centralised technologies.

Opposition to technology

Opposition to technology is not new. There are inevitably fears that new technologies will result in major social or environmental dislocations and sometimes this has proved to be the case. However, the scale of opposition to wind farms has, so far, been trivial in comparison with the opposition that has emerged to nuclear power. The same is true for the other renewables. Even so opposition does exist and in part this is because renewable energy systems tend to be very visible.

This has been a particular problem for wind turbines and the problem has not been restricted to the UK. For example, during the initial 'wind rush' in California large numbers of sometimes poorly designed machines were installed near to major highways, so that they were seen regularly by many people. In *Wind Power Comes of Age* Paul Gipe notes that the developers 'could not have picked a worse place to begin than the Altamont Pass' since the meteorological conditions meant that 'for much of the year, and especially during the morning commute, even the best turbines in the best locations stand idle. And all the early trouble prone turbines were installed immediately adjacent to Interstate 580, one of the busiest highways in the state, with 36 million vehicles passing by per year' (Gipe 1995: 275).

Although, Gipe claims, well-designed machines on good sites might be expected to operate 50 to 75 per cent of the time, this still meant that there would be times when they were becalmed or out of service for maintenance. The very visible spectacle of inoperative machines suggested to passers by that they were an inefficient source of power.

However, leaving visual effects such as these aside and assuming that noise problems are avoided, wind turbines have few other impacts compared to other renewables. Perhaps the worst case has been waste-into-energy plants, involving power production from the combustion of domestic and industrial waste. Some of these projects have met with strong local opposition from residents concerned about toxic emissions and unconvinced by assurances that these would be kept within controlled limits.

Similarly there has been opposition to geothermal projects, some of which have resulted in the emission of noxious fumes. In Paris in 1980 there were objections to the noise during the drilling phase: local residents put up a poster saying 'Oui à la géothermal, non aux nuisances' More seriously, in Hawaii there were bitter conflicts over geothermal projects, fuelled in part by local resentment at what was seen by some people as the desecration of their natural heritage.

In the cases of both waste and geothermal projects, local opposition has sometimes resulted in their abandonment. This has been a serious problem in the UK for some waste combustion projects supported under the Non-Fossil Fuel Obligation (NFFO): only around 72 MW of the 311 MW of municipal and industrial waste combustion capacity supported under the first two NFFO rounds was successfully commissioned.

As we have seen, some wind farm projects in the UK have also had to be abandoned following local opposition and adverse planning decisions. However, the bulk have been accepted. For example, as has already been noted, most of the initial series of wind farm projects supported under the first NFFO got through and, although objections began to emerge subsequently, 54 MW of the 84 MW of wind capacity contracted for under the second NFFO has been successfully commissioned. And eleven wind farm projects supported under the third round of the NFFO obtained planning permission, while only seven were turned down.

The economic situation for wind projects has been enhanced by continued technological development, improved operating experience and the removal of the 1988 deadline for the levy for the third NFFO and subsequent NFFOs. The third NFFO order, announced in November 1994, involved contracts of up to fifteen years and offered an average contract price of 4.32 p/kWh. A total of 165 MW of wind farm capacity was contracted for, including 20 MW of smaller wind projects (of less than 1.6 MW). Given that the economics of wind power have improved, continued expansion seems likely: 227 wind farm project proposals, representing 1,461 MW (DNC), were put forward for inclusion in the fourth NFFO, scheduled to be formally set in early 1997. Although obviously not all of these proposals will be taken up, clearly wind power is well underway in the UK.

Even so, getting acceptance is still not easy. To explore some of the issues concerning public responses to renewables, it is worth reviewing the lessons from our wind farm case study, by looking at how the debate over wind farms developed in the UK following the initial flurry of projects and reactions described so far.

UK wind farm development: the next phase

As we have seen, the wind farm issue came to a head in the UK in the early 1990s. Subsequently, although objections have continued, the debate was to some extent set in a more productive vein by the extensive review carried out by the all-party House of Commons Welsh Affairs Select Committee in 1993–4. This took evidence from all the protagonists. The Committee concluded that as long as they were sensibly planned, wind farms could be acceptable in Wales, and in effect rejected some of the more aggressive claims of the objectors (Welsh Affairs Committee 1994).

The subsequent production of *Siting Guidelines* by the Friends of the Earth (FoE 1994) and *Best Practice Guidelines* (British Wind Energy Association 1995) also helped to clarify the situation, as did the publication of further local opinion surveys showing overall support.

However, the environmental point had been made and subsequently the wind developers seem to have avoided complacency. Instead, what has emerged is a recognition of the need to negotiate between the various interests, in terms not too dissimilar from those presented in the model of interactions in Chapter 1, so as to balance costs and benefits more equitably.

Certainly the issues are now more strategic and less partisan, with environmental concerns being taken seriously. Some wind power supporters argue that large wind farms in remote areas are one way to achieve a significant total wind power contribution, while avoiding local objections. Others, however, suggest that smaller projects are preferable, as indeed already seems to be indicated by the success of, for example, the Windcluster and Farm Power and Waste wind farm developers in obtaining sites for smaller projects.

Community ownership

It has also been suggested that community initiated and owned projects would be more successfully accepted – following the example of the local wind co-operatives in Denmark. The argument is that if wind projects involved some form of local control and economic benefits then the impacts might be judged more favourably and, by implication, that absentee owners and backers will inevitably be less sensitive to local concerns.

Of course, even with conventional commercial projects there are some local benefits in terms of employment during construction and tourism subsequently. Some sites have set up visitors' centres which have proved very popular, injecting cash into the local service economy. Local farmers can also benefit, by renting out fields for use by wind farm developers (per annum rentals of £1,000–2,000 per acre or per turbine are typical), although this can sometimes lead to conflicts and resentment if residents feel that they are adversely effected by the wind farm.

In some cases wind farm developers have offered packages of 'community support measures'. For example, National Wind Power has provided £100,000 for a charitable trust to be used to support local schools, colleges, students, apprentices and training schemes in the area around its wind farm at Bryntitli in mid-Wales. It has also provided a £5,000 p.a. fund to support local environmental improvements in the area of the wind farm and has set up a community fund which will receive £5,000 p.a. for the benefit of local inhabitants (NATTA 1994c).

Schemes like these might give local communities a vested interest in the project or, failing that, might be seen as compensation for any disbenefits, although equally they might be seen by opponents of the project as a form of bribery.

In principle, local involvement could range from the purely token, right through to full-scale ownership. However, in the UK economic environment the latter seems a little unlikely at present on any significant scale. More probably there will be continued conventional corporate developments, softened perhaps with some attempts at local level involvement. In which case there are also likely to be continued allegations of corporate insensitivity and 'profiteering', with the wind programme being seen, at least by objectors, as being driven by profit-oriented commercial interests.

As has already been indicated, this may be unfair, not simply because the profit margins are in fact very tight, but because many of the developers have strong environmental commitments. However, from the grassroots NIMBY viewpoint, this may not be convincing. Sensitive local consultation, well in advance, is an obvious priority, but this takes time.

Probably the key issue will therefore be the pace of any future programme – as well as, of course, its scale. If a large and rapidly expanding programme is seen as driven purely by profit concerns, then opposition is bound to grow. If a commitment to wider social and environmental goals can be convincingly displayed, then perhaps more support will be forthcoming; many people are concerned about the greenhouse effect and acid rain. In order to engage with that concern, the wind power lobby will have to win the argument that it can make a significant contribution.

Issues for the future

Although the bulk of wind farm proposals in the UK have so far been successful, the advent of local opposition in some areas has slowed progress. Nevertheless, even in a crowded island like the UK, the prognosis for the future of wind power looks good, as long as the developers remain sensitive to local concerns.

The long-term technical potential for on-land wind power is usually put at up to 20 per cent of UK electricity requirements, with basic siting constraints perhaps reducing this to 10 per cent. These are clearly only generalised estimates: much will depend in practice on specific local siting issues, as well as on the technology, the economics, and developments in the energy supply and demand relationship. For example, newer, quieter, cheaper machines are now becoming available (e.g. variable speed turbines) and operating experience should improve system performance. Certainly a key priority should be the development of low noise machines.

The introduction of cheaper, more efficient variable speed wind turbines could reduce pressure on upland sites, although to ensure that this translates into actuality would probably require something more coherent than the UK's current rather ad hoc approach to planning, which, as we have seen, operates on a site-by-site basis.

Obviously it is sensible to assess each project on its merits, with local issues and concerns being taken into account. But equally it could be argued that overly commercial pressures have to be resisted. Wind farm developers will presumably want to avoid expensive planning delays and one solution would be to develop some form of zoning, with areas suitable for development identified in advance, as in Denmark. Alternatively, and perhaps more aggressively, siting height or even maximum average wind speed limits could be imposed to protect upland sites, perhaps with exemptions allowed for areas where hilltop siting was not invasive.

The need for a consensus

Regulatory interventions like these may be one of the outcomes of the debate on wind farms. Certainly some sort of resolution of local conflicts seems necessary and, ideally, the negotiation of some sort of consensus on how wind farms should be developed. As NATTA, an independent national renewable energy network, stated in its evidence to the Welsh Affairs Committee hearings on wind power, the key issues would seem to be:

- *What is the 'carrying capacity' for wind farms in the relevant UK contexts?* Gradually some sort of consensus should emerge which could revise the overall estimates for realistic potentials.
- *What scale of deployment is best suited to the UK context?* Currently the basic options seem to be large wind farms in isolated areas funded externally, versus small perhaps partly or wholly locally initiated/owned projects, although a range of possibilities exist in between, depending on the availability of specific sites (e.g. marginal land).
- *What form of public consultation is most appropriate?* The local

> planning system has tried to cope with this novel technology, but the pace of development is sometimes seen as too fast to allow for lessons to be learnt and 'case law' developed and digested. And there is the perennial problem of what to do about intractable objecting minorities.

NATTA, which has been monitoring the wind power debate since the start of the programme, goes on to argue that:

> The process of reaching a consensus on issues like this is not helped by the polarisation of opinions into simple 'pro' and 'anti' camps. It is of course understandable that local residents faced with, as they see it, potentially threatening projects, will resist. Equally, it is understandable that developers and pro-renewable environmentalists will become impatient. And some developers have perhaps been insensitive as a result. However, if renewable energy technologies are to play their part, for example, in reducing the emission of greenhouse gases, then we will need a less confrontational approach.
>
> (NATTA 1993b)

To reduce the potential for conflicts and provide guidance for local planners, NATTA argues that central government could and should play a key role in defining national energy policy and the role of renewables within it. However, as we have seen, at present some of the responsibility for this seems to have fallen to local councils and environmental organisations. The arena of public debate is obviously an important one, although, as NATTA comments: 'It is one that can be influenced by unrepresentative pressure groups. Those with more formal responsibilities must obviously try to gather accurate data on actual public opinion.'

In addition there would seem to be an urgent need for independent studies of local opinion, to help inform the debate. Failing that, as NATTA put it: 'The wind farm debate, and, possibly, others in the same field, may end up reflecting unrepresentative views and vested interests, either pro or anti. This would be very unfortunate if we are to try to develop, or negotiate, a consensus on how to proceed in future.'

Conclusion

As we have seen, the wind farm debate became quite strongly polarised, although it is difficult to judge the significance of the opposition. The pro-wind lobby was confident that as more people saw real wind farms, rather than reading perhaps misleading press reports about them, support would grow. At the same time, the anti-wind lobby became increasingly confident, given its success at gaining media access and political attention. However, despite this, the Country Guardians' call for a moratorium on wind power was turned down by the Welsh Affairs Committee and the objectors have subsequently had to be content with opposing projects one by one.

Although conflict continues at local level, the national level debate over wind power has become somewhat less strident. This, as has already been argued, is a welcome development. For in the end, unless a stand-off is maintained, the debate must lead to some form of negotiation on the extent to which wind farms can be introduced in the UK countryside, which means that some of the conflicts between interest groups will have to be resolved.

It is relatively easy to depict objectors as simply responding to NIMBY sentiments. But they can also reflect an important concern for environmental protection which must not be lost in the wider strategic debate over the role of renewables in the UK energy context, e.g. concerning which technologies should be emphasised and how rapidly they should be developed.

As already indicated, an important issue in this debate is the question of how fast the various options could be developed and to what extent. For example, given the inevitable limits on land deployment, offshore siting is an attractive option. The energy resources offshore are usually very large and the environmental impacts associated with tapping them are generally low. Offshore wind farms have already been developed off the coast of Denmark, the Netherlands and Sweden. However, despite the large energy potential there are, as yet, no plans for offshore wind farms in the UK; on-land siting is still the cheapest option.

In this situation a constructive public debate is needed on wind farms, which could also lay the basis for the sound development of other renewables. This forum is important since the renewables represent, essentially, a new form of energy technology, with new types of impact. Whereas previously the accent has been on concentrated energy sources and centralised power plants, the trend seems to be towards the more decentralised use of diffuse natural energy flows and sources. The emphasis would thus be on natural processes occurring in 'real time', not with inherited wealth from stored fossil or fissile energy. This opens up a whole new range of planning and land use issues which have only just begun to be discussed.

As yet, there are few techniques of analysis to help us assess the relative importance of any social and environmental impacts. The use of cost benefit analysis, in which an attempt is made to give economic values to costs and benefits, seems unlikely to provide anything more than a very partial means of assessing renewables. New techniques seem necessary, for example reflecting the impact of extracting energy from diffuse natural energy flows, perhaps following the approach being developed by Clarke as discussed in Chapter 3. There is also a need for more detailed studies of how public awareness and understanding develops and changes. And there is a need for new methods of social negotiation to resolve the inevitable social conflicts.

One practical possibility is to develop the UK Environmental Council's 'Environmental Resolve' initiative. This involves training workshops in conflict resolution techniques, designed for planners, developers, environmental groups and so on. The British Wind Energy Association made use of a 'consensus building' approach, involving discussions and negotiations with all the key parties, including groups opposed to wind power, as part of the process of drawing up its best practice planning guidelines (BWEA 1995). Consensus building approaches are becoming increasingly common in the renewable energy planning field (Hyam 1995). At the same time there is also a need for more general social mechanisms for consultation and conflict resolution.

More information on the issues needs to be disseminated to the general public, in order to improve the quality of the debate. In Denmark the national referendum on whether to embark on a nuclear power programme was preceded by a major national education effort, via local meetings, seminars, and so on. Environmental education generally seems a vital requisite for informed debate on how best to develop sustainable energy technologies.

While some renewable energy enthusiasts have expressed concern at the tone of some of the current debate in the UK over the merits of wind farms, in general public discussion should be welcomed, as long as it is as well informed as possible. Renewable energy technology is meant to be both socially and environmentally acceptable and there is a need for public debate over specific projects and over longer term development patterns. This is part of the process of bringing technology under more direct social control. Of course, as with all exercises in democracy, there can be costs, not least in terms of delays. To try to limit these there may be a need to develop new ways of conflict resolution via an extension of public consultation and participation in decision-making and planning processes.

In principle the development of renewable energy should be more amenable to local social control since many of the technologies are relatively small-scale, and their nature, function and likely impacts are fairly easy to understand. We will be returning to look at some of the specific issues that might form part of the debate over how best to develop renewables in Part 4.

To summarise, our case study of the pattern of initial opinion formation highlights the problems of bias, prejudgement and conflicting social and environmental priorities. These problems are not unique to wind power but it is hoped that they may prove somewhat less intractable, given the 'transparent' nature of the technology and its impacts. What is needed is a constructive, well-informed debate on how best to develop renewable energy technologies like wind farms as part of a process of bringing technology under social control. This would seem to be a vital prerequisite for any attempt to move towards a more sustainable energy system.

Summary points

- Local planning and land use issues may well determine the scale and pace of renewable energy development.
- Some objections may have been based on misinformation.
- But not all objections are wrong: there is a need for careful assessment and negotiation to avoid problems.
- Well-informed criticisms might be seen as being part of an attempt to bring technology under social control and as part of the process of moving towards a sustainable society.

Further reading

The report by the Council for the Protection of Rural England, *Renewable Energy in the UK* (1995, CPRE) provides a useful overview of the strategic debate in the UK, stressing the need to balance environmental and commercial concerns.

ETSU, the DTI's Energy Technology Support Unit at Harwell, can provide guides for planners and developers covering most types of renewables.

A guide to some of the research on public reactions to renewables in the UK and USA was included at the end of Chapter 10.

Part 4 Sustainable society

In Part 4 the emphasis moves from specific sustainable energy technologies and their problems and on to the wider issue of the prospects for sustainable development generally. It asks whether technical fixes will suffice or whether there will be a need for more radical changes. Will the use of sustainable energy technologies allow economic growth to continue, or are there ultimate environmental limits to human aspirations and expectations?

By way of conclusion, Part 4 also looks at some of the ways in which the conflicting interests of people and planet might be resolved and at the idea of 'thinking globally and acting locally'.

13 Sustainable development

- **The limits of technical fixes**
- **The need for social change**
- **Incentives for change**
- **Strategic choices for the future**

Given that we live on a small planet with finite natural resources and a fragile ecosystem, environmental sustainability may be incompatible with continued economic growth, at least of the current sort. But as this chapter shows, not everyone agrees with this proposition: the optimists believe that human ingenuity and technical fixes will suffice, the pessimists believe that more radical technical, and perhaps social, changes may have to be made. Radical changes may appear threatening to those whose livelihood is linked to the present arrangements but, as we shall see, there may also be some benefits.

Strategic issues for sustainable development

The key message of our analysis so far is that technology presents problems, but also possibly some solutions, even if these may have problems of their own. These problems require a process of social negotiation – as part of the larger process of moving towards a sustainable future.

Many strategic and tactical issues emerge from our discussion so far. Some are essentially 'technical', although they may have major social implications, e.g. should renewables be developed on a local, decentralised basis or must they be integrated on a larger scale? We will be looking at this sort of issue in Chapter 15.

Our focus in this chapter is on strategic issues which take us beyond the technical and raise wider social and political questions, concerning the overall direction not just of technology but of society.

Following on from the criteria developed in Chapter 3 and the discussions in subsequent chapters, the major strategic questions for the future in the energy field would seem to be: to what extent can renewables and conservation help us move towards environmental sustainability, and how fast can and should renewables and other 'green' technologies be developed and deployed to this end?

These may appear, initially, to be relatively straightforward 'technical' questions, concerning the ability of technology to deliver the required amounts of energy. However, they raise broader issues, for example, concerning exactly what is meant by sustainabilty. Perhaps the hardest question concerns the relationship of sustainability to economic growth. Renewables and conservation may be able to sustain economic growth up to a point, but is continued economic growth of the current sort viable or even desirable? A linked but slightly more tractable question is whether technical fixes will suffice or will there also be a need for more general social change?

Technical fixes succeed

From the purely technical fix point of view, a move towards sustainability, in terms of significantly reduced levels of resource use and resultant emissions, seems likely to be technically feasible. Pollution can be cut dramatically by 'end-of-pipe' measures as well by the introduction of more radical 'clean technology'. Much has already been done: for example, the Business Council for Sustainable Development has noted that between 1970 and 1987 the West German chemical industry managed to cut emissions of heavy metals by 60 to 90 per cent while boosting output by 50 per cent, and the Nippon Steel Corporation cut emissions of sulphur dioxide by 75 per cent. In the USA Monsanto pledged to cut air emissions of certain hazardous chemicals by 90 per cent by 1990, en route to a target of zero emissions; and in general industry seems confident that it can reduce its impact to negligible levels, given time (Schmidheiny 1992).

Faced with successes like these, it would seem that human ingenuity can solve just about all the environmental problems. Technology and technical fixes can provide the answers. One of the most famous adherents to this view was Hermann Kahn, the US futurologist, who was critical of what he depicted as the pessimism of the environmentalists. While some environmentalists in the 1970s were drawing up doom-laden scenarios of imminent energy shortages and environmental collapse, Kahn and his followers pointed to technological breakthroughs which could resolve some of these problems. In the short term Kahn proved right: the expected crisis went away. In the environmentalists' view, however, it had only been postponed, not resolved. Kahn was unconvinced. His last book, *The Resourceful Earth*, summed up his views: the planet was immensely rich and mankind was clever and could develop technology to resolve most problems (Kahn and Simon 1985).

The technical fix position adopted by Kahn and other similar optimists, is sometimes associated with what has been called the 'cornucopian' view, named after the legendary Greek goddess of plenty. The cornucopians hold that abundance is there for the taking since, unlike other animals, human beings have the intelligence to create technologies to meet their ever-expanding needs and to do so in ways which will not undermine their continued enjoyment of life and the environment. At the extreme, this view can be used to justify an assertion that environmental problems are not very important. These problems can be 'fixed' and there is no need for heavy intervention in economic affairs: certainly nothing should be allowed to slow economic growth. Indeed it was growth which could provide support for the development of new cleaner technologies, should they be needed.

This sort of analysis is often heard on the political right where it is held that there must be no interference with free market competition. The approach has been well represented by Wilfred Beckerman's book *Small is Stupid*, which mounts a serious attack on current green thinking (Beckerman 1995).

This chapter is not the place to enter into what is obviously an ideological debate. Suffice it to say that more liberal-minded critics feel that some form of intervention is vital. The real debate is over the extent of the intervention and how precisely it should be carried out. Modern liberal-minded environmentalists, like UK economist David Pearce, argue that if the market is given the right signals it can be used to bring about appropriate behaviour by investors and consumers. Markets can be shaped by a whole range of fiscal measures such as taxes, and by state subsidies, grants, regulations and controls, in order to prioritise more sustainable developments (Pearce 1992). Increasingly the substance of debates over modern politics and economic policy in the developed countries concerns exactly how such mechanisms should be used and precisely what the correct balance should be.

However, the key strategic questions still remain. Whatever the mechanism used to stimulate them, can technical fixes really solve our environmental problems or will there also have to be social changes? And even if technical fixes can resolve most of our current problems, can growth continue indefinitely?

Beyond technical fixes

To try to get to grips with these questions, let us go back to the basic issue of energy supplies. As we have seen, energy wastage can be reduced dramatically in most sectors of use (domestic, industrial, etc.) by the introduction of energy conservation techniques and the use of energy efficient devices, products and production systems, making it easier for renewables to supply the remaining energy requirements. Savings of at least 50 per cent and perhaps up to 70 per cent or more are seen as technically possible. As Figure 10.1 illustrated, in principle renewables could then meet around 50 per cent of global energy requirements by 2040 or so.

The study produced for Greenpeace looked even further ahead (see Figure 13.1). Greenpeace estimated that renewables could supply nearly 100 per cent by the year 2100, given proper attention to energy conservation. So there is no 'doom and gloom' here. Indeed Greenpeace claimed that this could be achieved without there being a need to curtail economic growth. In their scenario, energy consumption increases dramatically. This is based on the assumption that the overall world gross

Figure 13.1 ***Greenpeace's fossil-free global energy scenario: primary energy supply***

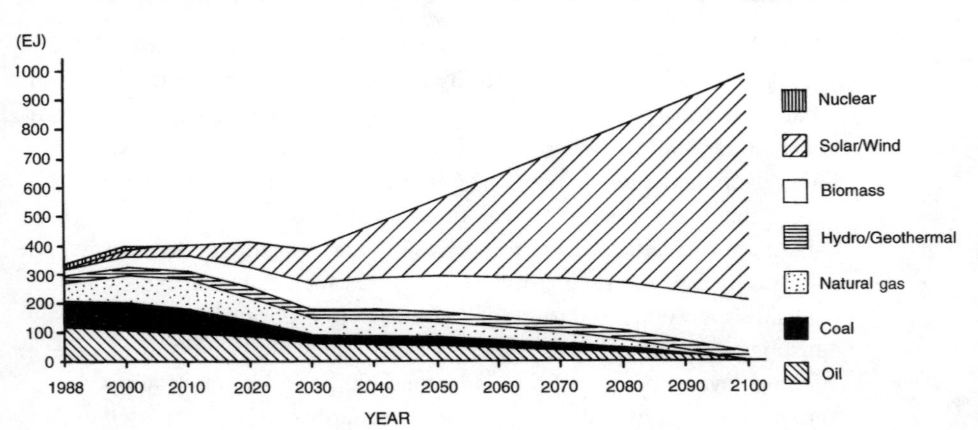

Source: Greenpeace, 1993.

national product continues to increase, although an attempt was made to factor in some redistribution between countries in the North and South (Greenpeace 1993).

Long-range scenarios like this are useful as a way of mapping out not what will happen, but what is technically feasible. The Greenpeace scenario is based on some fairly optimistic assumptions about the successful development and deployment of new renewable energy technologies. For example, it assumes fairly widespread use of photovoltaics, hydrogen-powered fuel cells and biofuels. But it does seem reasonably realistic, especially since it does not exhaust all the technical possibilities: for example, it does not assume any contributions from wave power or tidal power.

The real issue, of course, apart from whether the specific technical options can deliver, is that of implementation. Greenpeace notes that the key factor in bringing this about is the political will to make the transition, with the implication that there would have to be serious commitment from and intervention by governments.

Obviously this is a radical scenario – and no one as yet has adopted a programme anything like it. But there are some signs of interest in making a start. Some countries seem to have concluded that environmental problems like global warming warrant a radical approach. For example, Denmark and the Netherlands have embarked on radical programmes of energy conservation, as well as the introduction of renewable energy technologies, backed up by a system of subsidies. These measures are designed to help meet some challenging targets for reducing emissions of, for example, carbon dioxide, a key 'greenhouse' gas. Denmark has already made a commitment to a 20 per cent cut by 2005.

Going even further, one pan-Scandinavian proposal was for up to 95 per cent reduction in carbon dioxide emissions by 2020. This was seen as the sort of level of reduction that the 'advanced' Scandinavian countries would have to make in order to get the world total down by around 60 per cent, given that the advanced countries historically have benefited from using the world's fossil fuel resources (Meyer *et al.* 1993).

This level of reduction might possibly be achieved via technical fixes, without having to reduce the actual level of energy consumption at the point of use. For example, the Greenpeace scenario assumes increased global end use of energy as well as increased primary energy use, while still achieving dramatic reductions in carbon emissions.

However, the Greenpeace scenario may be optimistic: it could be that the various technical fixes will not allow overall levels of growth in energy use to increase in this way, at least not everywhere in the world. As we noted, Greenpeace assumed a degree of economic redistribution between the North and the South and, presumably, lower rates of growth in energy use in the advanced countries, but this may be optimistic. Even given technical fixes, there could be a need for more than just a slowing of growth rates to achieve a sustainable balance. There might be a need for overall reduction in the absolute levels of material consumption, particularly in the advanced countries. At the very least there could be a need for overall patterns and levels of consumption to change, with perhaps a move away from the Western emphasis on 'quantity of material consumption' to 'quality of life'.

The idea that consumption might be voluntary limited is gaining ground. There are already 'sustainable consumption' initiatives in Scandinavia, where it is sometimes argued that consumption levels are in any case reaching saturation (Norgard *et al.* 1994). There has been some debate in Scandinavia and the USA over the idea of 'voluntary simplicity', that is consciously choosing a less consumerist approach to life. Similar views have emerged elsewhere. For example, in Australia there is a strong environmental movement, the radical wing of which seems convinced that technical fixes will not be enough, and that there will also be a need for major social changes (Trainer 1995).

Some greens believe that social and technological change is not only environmentally necessary but also morally and socially desirable, given the vast imbalance of affluence around the world. This amounts to a call for a new set of environmental and global values. In some cases this can be taken to imply a radical change in lifestyle, new ways of thinking and more co-operative ways of organising to meet needs. For example, the 'deep ecology' view, which began to emerge in the mid-1980s, stresses the need for human beings to get back in touch with nature, and for a new emphasis on the spiritual as opposed to the material (Devall and Sessions 1985). This approach is often two-edged. On one hand, the deep ecologists claim that living in this way could actually be more fulfilling but, on the other, they insist that human beings have no right to live in any other way.

An end to growth?

Underlying many of the deep ecology critiques of contemporary industrial society that have emerged from the radical end of the green movement is the idea that economic growth, at least of the type experienced so far, cannot be continued. For example, it is sometimes argued by deep ecologists that even if massive energy conservation programmes were to be introduced, toxic emissions cleaned up and a switch to renewables achieved, the planet's ecosystem will still be overwhelmed if the ever-growing world population attempts to attain Western levels of material affluence. The only alternative, on this view, is to move toward more sustainable lifestyle patterns, based on drastically reduced levels of consumption: essentially 'living better with less'.

This is not a new idea: the concept of 'voluntary simplicity' as a response to environmental problems has been around since the 1970s (Elgin and Mitchel 1977). However, it is based on a prognosis which Kahn and the cornucopians claim is unduly pessimistic.

There is no way in which this debate between optimists and pessimists can be easily resolved, at least not within the context of this book. The discussion is usually couched in technical and economic terms, with much use of statistical prediction of trends. Quantitative analysis can certainly help to try to identify absolute limits (e.g. in relation to toxic emissions or greenhouse gas concentrations) and statistical patterns (e.g. in population growth or increases in economic productivity). We have already mentioned the idea of an overall energy limit in Chapter 3, and in the next chapter we will be looking at some of the overall constraints on the pattern of global energy use and industrial development.

In reality, however, much depends on assumptions and beliefs concerning what is possible and what is desirable – and these are not always susceptible to objective analysis. Even some of the terms and concepts used can present problems. This is nowhere more clear than in the way the term 'sustainability' has been interpreted and used. For radical greens its meaning is clear: the long-term survival of the planetary ecosystem, which many hold is incompatible with continued economic and industrial growth. At the same time optimists sometimes talk of 'sustainable growth' or even 'sustained growth'. The term sustainability has become not an absolute but more of an adjective, describing an attempt to move in the direction of more environmentally compatible approaches.

While technical analysis may help reduce some of the uncertainty, in the end the debate over sustainability and the appropriate means of attaining it would seem in essence to be political and ideological, concerning, for example, conflicting social priorities, generalised environmental ethics and vested economic interests. If this is the case, in practice the debate can only really be resolved by political means – by the messy process of reaching agreements, among people with differing views, interests and amounts of influence, on what should be done.

If the pessimists are right, there would be a need for what could be perceived as painful social changes, accompanied by some hesitancy. For the cornucopian optimists are not the only ones who would find unwelcome the proposition that lifestyle patterns and consumption levels will have to change. Many people in the advanced countries would no doubt see any change in their material living standards as a serious imposition and would resist it. They may be happy to engage in recycling or to purchase environmentally friendly products, but many would no doubt perceive actual net reductions in material consumption as a threat to their standard of living and an unwanted return to frugality. A shift away from consumerism and any suggestion of lowered economic growth rates would also no doubt be resisted by industrial and commercial interests, which might fear that profits would be undermined.

We are back to the conflicts outlined in the model of interests presented in Chapter 1. In this case consumers and producers share a perceived common interest in continued economic growth. This is as true in the developing world as it is in the developed countries: people in the developing world would like access to similar benefits as those enjoyed in the developed countries. It is therefore not surprising that technical fixes are popular: they seem to offer the possibility of the benefits of continued growth without the environmental problems.

Even so, there is evidence that there is significant generalised support for more radical change, as indicated by Table 13.1, based on a study by Milbrath of public opinion in the USA, UK and Germany (Mann 1981: 57). More recent studies, for example, reported by the UK Central Statistics Office, have confirmed a trend in concern over environmental matters (CSO 1992). However, whether these generalised concerns could be translated into acceptance of real reductions in consumption levels remains to be tested. Voluntary shifts in consumption patterns may occur, as people become more aware of environmental problems, but shifts in attitude and behaviour are more likely to take place if there are some clear benefits. Therefore it may be that consumers are likely to need positive incentives before they make major lifestyle changes.

Incentives for change

Individuals find it difficult to accept change in lifestyles and it seems likely to be even harder for nations to adopt new approaches which may go against short-term vested interests. As the slow progress on

Table 13.1 ***Public opinion on environmental protection***

Preferences for economic growth (score 7) or environmental protection (score 1)

USA	
General public	2.99
Environmentalists	1.97
Business leaders	4.16
England	
General public	2.80
Conservation society	1.48
Business leaders	3.38
Germany	
General public	2.99

Source: Milbrath in Mann 1981

responding to global warming indicates, there is still some way to go before the global community will be able to respond to the serious environmental challenges that many feel lie ahead. Vested interests in the continued use of existing energy sources still dominate: for example, some oil-producing countries have resisted attempts to develop international obligations concerning reductions in carbon dioxide emissions.

In order to try to break out of this sort of impasse there is clearly a need for concerted efforts at negotiation and commitment to action at all levels, stressing communality of interests and trying to resolve sectional differences. After all, in the longer term, all the world's peoples share a common interest in the health of the planet. A shift to a more sustainable approach will involve significant restructuring of industrial activities but, as we shall see, there may actually be some longer term commercial advantages. More generally, any disbenefits associated with a shift to sustainability should be set against the overall shared benefits. This, of course, is not always easy because of the conflicts of interest that already exist between the various vested interests within any particular country and the competing concerns of countries themselves.

Reaching international agreements on change is likely to prove particularly difficult, especially if the implication is that economic growth and competitiveness will be affected. Vested interests in the status quo remain strong. For many radically minded people, it is unlikely that agreements can be reached when the world economy is dominated by powerful multinational companies which are seen as having no concern for the fate of individual countries. In the same way, it is also often regarded as unlikely that agreements can be expected between the North and South when many of the developing countries are trapped in debt relationships with the rich countries.

From this viewpoint, there will have to be major social, economic and political changes before there can be an effective international response to global environmental problems. At the very least, international economic activities must be subject to some form of control to ensure fairer trading relationships and reduce exploitation of both people and the planet. Obviously this opens up issues that are well beyond the remit of this book. We will be looking at the overall prospects for economic expansion worldwide in the next chapter, where we also discuss the need for technology transfer and aid programmes. Let it now suffice to say that there is a clear need for changed priorities, not least by multinational companies.

Of course it could be that the major companies of the world may introduce some changes to the ways in which they operate, not so much on an altruistic basis but in order to maintain their longer term commercial survival. On this view, since the world's environmental problems will not disappear, and pressures to clean up are likely to increase, companies and countries that invest now in 'clean green technology' will be better placed competitively in the future. Of course, some companies will be able to ignore these problems in the short term and when and if problems emerge they will move on to other activities and/or countries where constraints are less. But in the longer term there will be no escape from environmental and resource limits and from competition by companies which have made the switch to more sustainable approaches.

As we have seen, some companies have already embarked on quite radical clean-up programmes, and this trend is likely to increase. Indeed it has been suggested that, given the prospects for economic growth and the need to stay ahead of the competition, the global multinational corporation may be one of the key agencies in bringing about a shift to more sustainable approaches (Wallace 1996). Certainly the market for pollution abatement and environmental clean-up technology is growing: it had reached £130 billion worldwide by 1990 and is expected to rise to £250 billion by the year 2000. Similar growth rates might be expected in the more radical 'green products/clean technology' and in the renewable energy fields.

However, even from the most optimistic view, there is a very long way to go. A shift to greener products and cleaner production processes and energy technologies will not necessarily resolve the inequalities in world trading patterns. Indeed the opposite might happen. As the early enthusiasts for alternative technology warned, it could be that the technology might be taken over for commercial purposes but the underlying social and political values might be ignored.

To be fair, some companies have responded to the idea of sustainable development in a wider sense by developing new approaches to trading with and investing in developing countries (Schmidheiny 1992). Unfortunately, they may be in a minority. More generally, far from sustainability being a major concern, many environmentalists would argue that the overall trend within the global economy is in the opposite direction, as a result of the drive towards 'free trade' around the world and the advent of a 'globalised' economy dominated by multinational companies (Korton 1996). Certainly, so far, environmental concerns have not figured significantly in the GATT free trade discussions. In the view of many environmentalists, there does seem to be a fundamental conflict between free trade and environmental protection – unless environmental controls can be imposed globally.

Even so, as we have seen, there are some signs of change in the way in which companies are approaching technology since some of them can see benefits from adopting more sustainable approaches. This at least is a start.

Green employment

While companies may gradually adopt a more sustainable approach for commercial reasons, or be pressured to by governments, there are some potential benefits associated with a shift to sustainable technology that may provide an incentive for wider support for change. For example, studies have suggested that at least as many new **jobs** could be created as would be lost due to the impact of tight emission controls on industry and the phasing out of environmentally undesirable technologies (Renner 1991; Jenkins and McLaren1994). Indeed it is claimed that not only could there be direct replacement, but there could also be a net overall gain in jobs. For example, a study by the UK consultants ECOTECH on the 'Potential contribution of renewable energy schemes to employment opportunities', carried out in 1995 for the UK government's Energy Technology Support Unit, concluded that by the year 2005 a total of 48,700 jobs would be created by the various renewable energy projects being supported by the NFFO subsidy, with a total of 32,300 jobs being created over and above the 16,400 jobs lost due to the replacement of conventional energy employment by employment on renewable energy projects (ECOTECH 1995). Similar studies have been carried out in the USA and, as Table 13.2 illustrates, the general conclusion is that renewable energy projects tend to create more direct employment than conventional energy projects.

Care has to be taken when making estimates of the job creation patterns associated with any form of investment. In addition to direct operational employment, on site, there are the jobs associated with manufacturing the hardware and indirect jobs elsewhere, providing the materials and

Table 13.2 ***Direct employment in electricity generation in the USA (jobs per thousand gigawatt-hours per year)***

Nuclear	100
Geothermal	112
Coal (including mining)	116
Solar Thermal	248
Wind	542

Source: C. Flavin and N. Lenssen, 'Beyond the petroleum age: designing a solar economy', Worldwatch Paper 100, Worldwatch Institute, Washington DC, December 1990.

components. The people in all these jobs will also create further jobs, 'induced' in the economy as a whole, when they spend their earnings. In terms of the conventional economic viewpoint, ultimately, therefore, taking the economy as a whole, the total number of jobs per £ or $ invested will end up being similar, regardless of the initial investment, although the dynamic pattern, location and type of job creation may differ. Put at its most severe, on this view, you will only get more net employment if you either invest more money or pay less wages, but you may be able to create more direct jobs quickly by investing in labour intensive technology.

There could still be some specific areas where there would be dislocations and a need for retraining. A shift to a sustainable future would clearly involve a need for careful planning to limit disruption. However, the areas of likely growth are those which are currently in decline. For example, aerospace workers could find employment in the wind power field, shipyard workers can work on wave energy technology and construction workers may find employment on tidal energy projects. Certainly, the changeover would involve a lot of new investment and this might be one way in which currently unemployed people could find work. The changeover might also provide more sustainable employment – less threatened by the economic ups and downs resulting from, for example, oil price variations.

Not surprisingly trade unions have shown interest in such possibilities. Indeed, some trade union groups were among the pioneers of the idea of shifting to 'socially and environmentally appropriate production', as witness the campaign mounted by trade unionists at Lucas Aerospace in the UK in the 1970s. They argued that rather than having to rely on defence-related production, job security could be better ensured by shifting to socially needed products and systems, and they developed their own plan outlining 150 new products which they felt they should work on. These included wind turbines, solar energy devices, and fuel cells. Several other UK trade union groups subsequently developed similar plans, as did some labour organisations in the USA, where a strong campaign for 'defence conversion' was mounted by workers in some parts of the defence industry (Wainwright and Elliott 1982).

These early initiatives were made at a time when the idea of sustainable technology, based on renewable energy and environmental protection, was in its infancy and in most cases company managements repressed or ignored the plans that emerged from their workforces. Indeed some of the UK activists were dismissed, in effect for challenging the management's right to choose what to produce. But times change and, while acceptance of workers' participation in decision-making generally remains low, some of the technical ideas that emerged from these campaigns are now more respectable – indeed many companies seem keen to be seen to be 'green'.

Trade unions in the UK have continued to press for environmentally sound production, although at times there have been conflicts over the future of nuclear power, with workers in the nuclear industry not surprisingly objecting to the generally anti-nuclear views of environmentalists. Even so, in 1986 the UK Trades Union Congress, which covers all unions, supported a moratorium on further nuclear development and in 1988 backed a nuclear phase-out motion (Elliott 1989).

Most of the UK trade unions became increasingly supportive of renewable energy and clean technology development as one way in which to secure safe and worthwhile employment (TUC 1989, NALGO 1990). In general, rather than being seen as leading to employment cutbacks, environmental policies are considered as one way to ensure sustainable employment (Tindle 1996).

This view also seems well established in the British Labour Party, which in 1994 made a strong commitment to environmental protection and sustainable development. Its 1994 environmental statement, 'In trust for tomorrow', set a target of a 10 per cent contribution to electricity supplies from renewable energy sources by 2010, and in February 1995 the Party promised to create 50,000 new jobs by an ambitious energy conservation programme.

While it is reasonable to expect job gains from new environmental programmes of this sort, in the longer term, when and if a sustainable energy and supply system were established, then overall employment might stabilise or even decrease, at least in the primary energy generation sector. It could be that there will be a compensatory shift to more work in the service sector, for example, on repair, renovation and recycling of energy efficient long-life products and systems, and a shift generally to shorter working hours. However, leaving long-term speculations aside, in the shorter term making the transition to a sustainable energy system would seem likely to involve a burst of employment. It could be that the 'jobs' issue could provide a way to resolve some of the conflicts outlined in the model described in Chapter 1, and in particular the conflicts between producers and environmentalists.

To summarise, the move to green technology and a sustainable energy future will involve many difficult challenges in terms of overcoming vested interests in the economic status quo. But far from being threatening, developing the necessary technology could have positive economic implications. If there is to be economic growth, this is one area where it could occur.

As we have seen, some companies have already shown an interest in developing clean technology and some governments are taking the issue of sustainability seriously. The next chapter widens the focus to look at what changes might be necessary, in terms of the overall global pattern of industrial and technological organisation, if sustainability is to be achieved.

Summary points

- Technical fixes may not be sufficient to achieve full global sustainability.
- There may also be a need for qualitative and quantitative changes in consumption patterns.
- Lifestyle changes and less reliance on economic growth may also be necessary.
- While making the transition would be difficult, a shift to sustainability may lead to improved quality of life and more secure employment.
- Employment gains may provide an incentive for making a shift to sustainable technology.
- Environmental issues are global, and the changes that are needed in order to attain sustainability are likely to have to occur on a global basis.

Further reading

Ted Trainer's book *The Conserver Society: Alternatives for Sustainability* (1995, Zed Press, London) is a useful corrective for those who feel that technology alone can solve our environmental problems. In similar vein, Ernest Braun's *Futile Progress*, (1995, Earthscan, London) argues that 'progress' as currently defined is an illusion, and that technology should be redirected to meet real needs rather than being driven ever more rapidly to create profits.

Adopting a somewhat different approach, *Changing Course: Global Business Perspective on Development and the Environment* by Stephen Schmidheiny and the Business Council for Sustainable Development (1992, MIT Press, Cambridge MA) argues that industrialists are already accepting the challenge to develop sustainable approaches, while David Wallace's *Sustainable Industrialisation* (1996, Earthscan, London) presents a forceful case for relying on international companies to restructure industrialisation on a global basis along more sustainable lines.

As a brief look at journals such as *Real World* and *Green Line* will indicate, not all environmentalists take this view. Earthscan and the Royal Institute of International Affairs have published a useful report on Sustainable development and the energy industries' (ed. Nicola Steen, 1994, London) which reviews some of the conflicting opinions of environmentalists and industrialists on the way ahead.

Nevertheless, environmental groups are keen to stress the benefits of a sustainable approach. For example, see the booklet by UK Friends of the Earth *Working Future: Jobs and the Environment* (1995, FoE, London). Finally, Earthscan, with the UN University Press, have published a collection of papers edited by Peter Hayes and Kirk Smith, *The Global Greenhouse Regime: Who Pays?* (1993) which asks how the responsibility for and the costs of environmental protection should be distributed.

14 The global perspective

- **Global economic development patterns**
- **The Industrialisation process**
- **Post-industrial society**
- **The potential for leapfrogging**

Environmental sustainability is a global issue: pollution does not respect national boundaries. The industrial system that creates it, and makes use of environmental resources, is increasingly organised on an international scale. This chapter explores the way in which industrial development patterns have influenced energy use around the world since the industrial revolution. It argues that new patterns of technological and industrial development may be needed in both the developed and the developing countries if the environmental limits that face the existing industrial system are to be overcome.

World development

Our analysis so far has suggested that, in very general terms, a sustainable future is technically feasible but socially and politically problematic, in that it might be seen as challenging the industrial and economic status quo. Our focus has been mainly on energy use in the developed countries, which historically have created and continue to create the bulk of the world's pollution. However, the developing world is catching up. Indeed, as Figure 14.1 illustrates, in terms of carbon dioxide production, the developing countries have collectively overtaken the developed world. Unless remedial actions are taken emissions are likely to increase dramatically.

Unless agreements can be negotiated on an international basis there is little hope for long-term environmental sustainability. However, this

Figure 14.1 ***Growth in carbon dioxide emissions from the advanced and developing world***

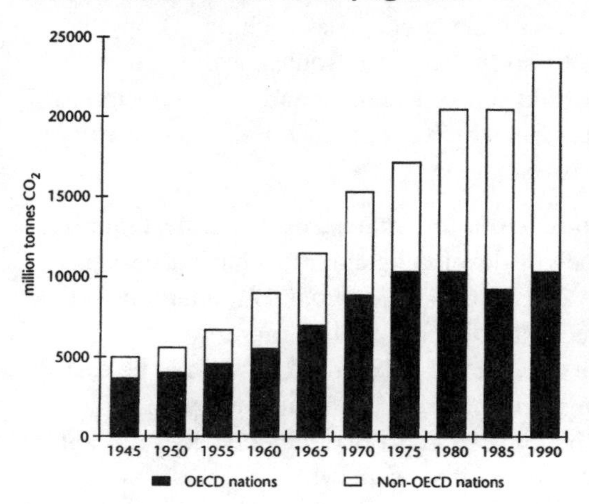

Source: Toke, 1995.

opens up the complex issue of global economic development and, in particular, the question not only of whether sustainable growth is possible, but also of the distribution of any growth between the rich and the poor of the world – the North and the South.

The developing countries of the South may see little reason why they should not follow the advanced countries path to industrial affluence, but can they and should they?

There has been much debate (e.g. following the 1992 UN Earth Summit in Rio) on how sustainable development can be achieved worldwide and on defining what that means for economic growth, trade and aid – not least in terms of transfer of technology from North to South. Renewable energy would seem to offer an important option. If advanced industrial countries can, given time, conceivably shift to renewables, then in principle it does not seem unreasonable for developing countries to follow that path, rather than copying the route which the West adopted historically. Renewables are well suited to many Third World countries, but whether it is possible for developing countries to 'leapfrog' over the dirty and inefficient intermediate stage of industrialisation remains to be seen.

A key factor shaping whether energy demand can be met in the future on a worldwide basis in an environmentally sound way will be the scale of population growth and the success of strategies for constraining it. However, this in turn depends significantly on the level of economic development that is possible: wealthier populations tend to have smaller families. This is not the place to explore the complex issue of population control. However, it does at least seem possible that if new energy technologies such as renewables can help developing countries to meet their energy needs without having to import expensive fuel or destroy their own environmental assets and rural economies, then perhaps the development process can create and distribute wealth more widely. Population growth may then be stabilised, this in turn making it possible for renewables to meet energy demands.

So far in practice within the developing world renewable energy technology has in many cases been relegated to a relatively minor role as an 'intermediate technology', with many countries seeking to copy the historical Western model of fossil fuelled or, in some cases, nuclear development. But given the right sort of technical and financial support, alternative routes may be feasible which are perhaps less economically draining and environmentally damaging.

There is always the temptation to opt for what seems the easiest, quickest and cheapest route and some key developing countries have access to substantial reserves of dirty fuel such as brown coal. But international agreements can be developed to try to ensure that a more environmentally sustainable approach is adopted. This will not be easy but in principle international agreements, if conformed to by all countries, can help ensure that the extra cost of adherence does not put participant countries at an economic disadvantage.

Not all countries will necessarily want or be able to subscribe to international agreements: this is matter for careful diplomatic negotiation. For example, in some cases international agreements are phased or regionalised to take account of different levels of economic development. Thus, as an interim measure, developing countries may be allowed to release more emissions than the developed countries, since the latter have not only produced the most pollution so far, but also have the technical and financial means (and, one might add, duty) to bring about improvements. At the same time, the international agreement may propose aid plans and technological transfer initiatives designed to help the less developed countries to initiate and use cleaner technologies.

To try to explore the way in which developing countries might move ahead, it may be helpful to look at the energy dimension in the pattern of world industrial development so far.

The post-industrial world

A gradual transfer of emphasis from energy and material intensive industries to an economy based on a reduction in such activities is typical of most mature industrial countries. Industrialisation involves a rapid increase in the use of energy in any given country in the process of building up the basic industrial infrastructure. The emphasis is typically on the extractive industries (e.g. coalmining), steel production and the manufacture of basic commodities.

Subsequently, however, as Riccardo Galli has put it in a study of 'Structural and institutional adjustments and the new technological cycle', as industrial countries mature 'quantity gives way to quality, light industries and services tend to dominate the economy, so that the need for materials and (energy) services declines' (Galli 1992: 781).

Industrialisation thus gives way to the development of a **post-industrial society**, with the emphasis less on primary 'heavy' industries and manufacturing and more on services, with communications technology playing a key role.

As part of this process, there is usually a shift away from the production and use of energy intensive **materials**. Although the production of basic materials continues elsewhere in the world, new materials are also developed which require less energy to fabricate and these are substituted for conventional materials in increasing numbers of products. The result is that the energy intensive production of materials like steel decreases, at least in developed economies such as the UK, USA and Japan. The process is also called **dematerialisation**. You may remember that this was the title of one of the Shell scenarios discussed in Chapter 8 (see Figure 8.3).

In some countries elements of this 'post-industrial' trend have been underway for some time. The UK was the first to industrialise, followed by Germany and the USA, and you would therefore expect each to reach the post-industrial phase in roughly the same sequence, although there might also be some variations in pace and even some 'leapfrogging'. For example, some latecomers to industrialisation might pass through the various stages more rapidly and perhaps overtake the initial leaders.

Of course, there may be other forces at work such as shorter term international trade patterns and longer term structural political and economic changes. But it seems clear that the contemporary process of dematerialisation and the shift to 'greener' technologies and 'cleaner' production systems which are underway in countries that have reached industrial maturity are at least in part an attempt to transcend the energy, resource and environmental limits of conventional industrial society. The result of this process could be a transition to a new 'post-industrial' pattern of organisation.

Energy trends

To try to get a feel for the process of industrial change, it is useful to look at the patterns of energy use of some key countries around the world.

They are shown in Figure 14.2, in terms of energy intensity, that is, the ratio of energy use to economic activity, the latter being measured in terms of gross national product (Farinelli 1994).

Figure 14.2 reveals some interesting trends. It seems that after industrialisation got underway the ratio of energy consumption to economic productivity in each country initially increased, then reached a peak and thereafter decreased, in the move towards a post-industrial economy. Moreover, the energy intensity peaks that were reached at some point after each country industrialised are all progressively lower. Each country in turn presumably learned from its predecessors and was able to make use of increasingly more efficient technology. Some of the key countries at the forefront of the new innovation cycle were able to use the new technologies to improve their energy efficiency.

The UK's energy intensity ratio (energy consumption/GNP) peaked at around 1880. Forty years or so later, around 1920, the USA reached a lower peak, although Germany had actually achieved a lower peak a little earlier. Forty years on again, around 1960, Japan reached an even lower peak, followed by Italy, at around half the UK's historical peak.

Figure 14.2 ***Energy intensity peaks for a range of countries as they have industrialised, in energy/GNP***

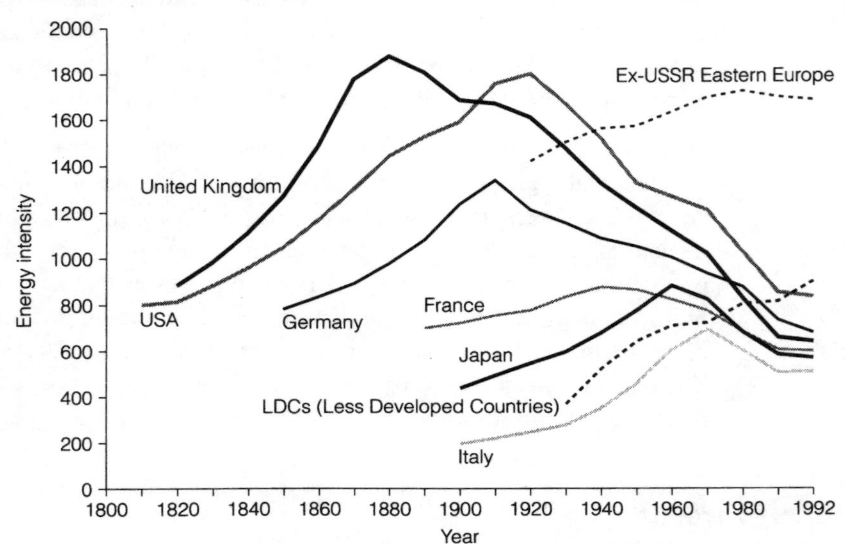

Source: Figure from *Renewable Energy*, Vol. 5, Part 1, Ugo Farinelli, 'A setting for the diffusion of renewable energies', pp. 77–82, Copyright 1994, with kind permission from Elsevier Science Ltd, Kidlington.

Kondratiev cycles

These patterns tie in roughly with the forty to fifty-year cycles of economic peaks and troughs first identified by the Soviet economist Kondratiev in the 1920s. Subsequent theorists in the West took up his ideas (especially after the economic crash, on cue, in the 1930s) and some have suggested that the key factor in creating the periodic booms was the development and adoption of a range of key new technologies (Schumpeter 1939). Theories vary as to precisely how this occurs, but in some versions of what is sometimes known as 'long wave' theory, the technological innovation process is supposed to be stimulated during economic depressions, when companies which had previously been able to expand markets for their existing products find that profits are falling. They therefore invest in new products and a boom results, after which markets begin to saturate once again so that the cycle repeats (Mensch 1979).

The first Kondratiev cycle is seen as having started with the industrial revolution, chiefly in the UK, based on coal/steam power related technology. The global economy is currently supposed to be at the end of the fourth Kondratiev phase and entering the fifth, with the main 'new technologies' being information and communication systems, coupled perhaps with biotechnology and renewable energy technology.

Certainly, in the current phase, one of the key factors shaping the development of new technologies would seem to be concern with the energy resource and environmental limits associated with existing energy technologies. Thus the technologies that emerge as part of the next cycle will, in effect, be those that transcend the limitations faced by the previous set of technologies. To an extent then the optimistic views associated with futurologists like Hermann Khan, as discussed in Chapter 13, would seem to hold some force. The limits to growth predicted by some environmentalists in the 1970s may in fact be lifted by a shift to a new set of cleaner and more energy efficient technologies.

However, we have to be careful not to see this as a simple mechanical prediction. Some current versions of long wave theory put the stress not so much on the advent of new technologies as on the social and cultural processes which surround their adoption (Tylecote 1991; Freeman 1992). As our case studies in earlier chapters illustrated, new technologies may not always be fully developed or widely accepted. There is a need for the right sort of social and institutional context and the process of social and institutional change may have a different timescale from that of technological change. These patterns of social and institutional change may of course reflect economic cycles, but equally there can be a mismatch between the economic cycles and the pattern of social and institutional change.

Not everyone accepts Kondratiev's long wave theory, or its predictions of regular cycles in economic activity, but it does seem clear that there are booms and slumps and that innovation does play a role in booms. It also seems possible that energy related developments are one factor shaping the new technological patterns. The cyclic pattern in energy intensity shown in Figure 14.2 certainly suggests that.

Of course, the matching of the energy intensity peaks in Figure 14.2 to the industrialisation process is not exact and in some cases is completely absent. For example, Italy, France and Japan were industrialised long before the energy intensity peaks shown in this figure. But in general there has been a similar trend in each country following on from industrialisation. At some point in the subsequent period new technologies are adopted which are more energy efficient.

Note that we are not talking about a reduction in energy use because of economic and industrial decline. Although this may happen in some cases, Figure 14.2 depicts an improved ratio of energy use to economic output, which implies economic success, but success attained by new means. Thus, typically, there is a shift from energy intensive 'smokestack' industry and mass production, with information based trading and service sector based activities becoming increasingly widespread, the latter generally less dependent on energy and material resources.

The most obvious exception is the ex-Soviet bloc, which seems to have peaked in energy terms in 1990 at a very high level (of inefficiency); it could be said that its industrialisation has not yet matured. A key issue for the future is whether the necessary technological and economic transition can be made.

The developing countries

What about the countries that have only just started their industrialisation process? Figure 14.2 suggests that the less developed countries (LDCs) are still on a rising energy intensity curve. The key issue in this context is whether they will be able the learn from the experience of other countries and make the necessary transition. As we have seen, most countries so far have been able to improve on their predecessors' performance. It would be tragic if the LDCs could not follow and instead replicated the worst excesses of the first countries to industrialise. It would obviously be preferable if they could 'leapfrog' the conventional industrialisation process, i.e. miss out on some of the intermediate stages and opt straight away for the 'cleaner', more efficient technologies.

Of course care must be taken not to read too much into the curves in Figure 14.2. For example, it could argued that GNP is not a very good measure of 'progress'. It might also be argued that part of the explanation for these curves is that the advanced countries shift to energy and material efficient service and manufacturing, and export the inefficient 'dirty' technologies to developing countries. The advanced countries certainly have more money to invest in cleaner, more efficient technology, whereas developing countries often have no choice but to use dirty inefficient technologies. It is also clear that industries in the developed countries sometimes 'dump' dirty inefficient technology on developing countries – to escape ever tightening environmental controls at home.

There would therefore seem to be a clear need for comprehensive global environmental controls to avoid this risk. In addition it would seem necessary to support 'technology transfer' to the developing countries, so that they too can move up the learning curve – and all nations can adopt clean, energy efficient, technology. Obviously this will be difficult to achieve. Some analysts fear that even if the developing countries are able and willing to adopt clean technologies, rapid economic growth could still lead to an awkward transition period during which overall global environmental impacts would continue to increase (Harper 1996).

Limits to growth

Even assuming that a transition to a more energy efficient, post-industrial, clean technology based future is successfully made by all developed and developing countries, the convergence zone on the bottom right of Figure 14.2 still implies global economic growth; and overall growth, albeit at lower levels, in total global energy and material use. Dematerialisation can only go so far. We are back to the familiar issue of the ultimate limits to growth. Even if the world does make the transition to the post-industrial economy, can this new more efficient, but still growing, level of economic activity be sustained on a global basis in environmental and resource terms? Or must an attempt be made to reach a steady state economy and a pattern of consumption which is not reliant on growth?

Switching to renewable energy sources would obviously help, but even so there are limits to what these sources could supply in total. The technological limit is defined by the total amount of energy that can be extracted from natural energy flows. As was noted earlier, Gustav Grob has called this the 'natural energy limit'. In terms of long wave theory, this might be seen as the limit for the next Kondratiev cycle, that is, the end of the fifth cycle, although it may take much longer than fifty years to reach it. Equally, even given the adoption of cleaner technologies, if economic growth continues to be the dominant concern, then the limit may be reached earlier.

Obviously we can only speculate on the longer term possibilities but, to summarise, it does seem that in the medium term renewable energy technologies, coupled with the use of information and communication technology and cleaner, greener and more energy efficient technology, can play a key role. They might help us transcend the energy and material limits to further development typical of the technological and economic arrangements of the current type of industrial society. However, there may be boundaries to subsequent growth, imposed by the limits of the renewable energy resources and possibly by other resource constraints. Technological innovation may help to raise the current environmental and resource limits to economic growth, but reliance on renewable energy technology would introduce new limits. In which case, at some point, a new type of technological and social transition would presumably be needed.

What next?

In this chapter we have looked at the influence of energy and resource related factors on technological and economic development. It has been suggested that environmental constraints, along with economic factors, may be the main influences shaping the pattern of technological development in most sectors of the global economy. However, the precise direction and the pace of change is difficult to predict. Economists like Freeman make use of the Kondratiev long wave theory, but have increasingly tended to downplay the idea that there are fixed fifty-year time periods. Instead, the emphasis is on 'phases' and transitions, and the adoption of new techno-economic paradigms, stimulated by social, economic and environmental pressures, without any specifically determined dates (Freeman 1992).

As has been suggested earlier, it may be that the next phase, the so-called 'fifth Kondratiev', will last longer than fifty years. For example, in purely technological terms it would take at least this long before renewable energy could make a major contribution to meeting the world's energy needs. Current energy predictions suggest that if a major commitment were made to developing renewable energy technology, it would take between fifty and one hundred years before renewables could supply half of the world's energy. So it could be some time before the 'natural energy limit' shown on Grob's chart (Figure 3.1) was reached.

Of course, it could be that well before the natural energy limit is reached other resource constraints will impose limits on economic activity that cannot be transcended just by technical fixes like the adoption of renewable energy technology. Technological innovation can certainly lead to economic growth but it may also, in the end, be unable to sustain it. In which case there might be a need for possibly painful social and economic choices.

However, the current task would seem to be to start on the long process of developing sustainable energy technology. In the next chapter, to begin to round off our discussion, we will look at the some of the more specific practical technological choices in the renewable energy field. This should highlight some of the constraints that lie ahead but also indicate some of the opportunities for positive change.

Summary points

- The Western industrialisation process cannot be repeated worldwide without major environmental consequences.
- Fortunately technical fixes and new technological development patterns may reduce this problem.
- However, there may still be longer term environmental and resource limits on the extent to which economic growth can continue.
- We are currently faced with a range of choices over how to develop sustainable energy technology, as part of the process of moving to a sustainable future.

Further reading

For a useful overview of possible trends, see Donnella Meadows, Dennis Meadows and Jorden Randers, *Beyond the Limits: Global Collapse or a Sustainable Future?* (1992, Earthscan Publications, London).

Kondratiev's long wave theory is complex, but if you wish to explore some of its implications see Andrew Tylecote's *The Long Wave in the World Economy* (1991, Routledge, London).

For a somewhat more pessimistic analysis of the process of technological development, see Ernest Braun's *Futile Progress* (1995, Earthscan, London). By contrast, Chris Freeman's *The Economics of Hope* (1992, Pinter, London) makes use of some of the ideas for long wave theory and presents a much more optimistic prognosis.

15 The sustainable future

- **Alternative paths to a sustainable future**
- **Choosing technology to fit**
- **Structuring the social control process**

Sustainable development is likely to require more than just technical fixes, but there is no one blueprint for a sustainable future. Instead there is a range of social and organisational as well as purely technological choices. This chapter argues that the process of choice must involve a process of negotiation over the social and environmental ends, as well as the technological means.

Social choices

The general conclusion that has been reached in the analysis in this book so far has been well expressed by Australian environmentalist Sharon Beder:

> So long as sustainable development is restricted to minimal low-cost adjustments that do not require value changes, institutional changes or any sort of radical cultural adjustment, the environment will continue to be degraded.
>
> (Beder 1994: 18)

We have looked at a variety of technical fixes and at some more radical technological developments. We have also discussed the view that there must also be significant social changes, or at least lifestyle changes. Patterns of consumption may have to change with at the very least more emphasis on conservation, recycling and less emphasis on consumerism. In the view of some greens more dramatic changes might also be necessary – for example, social, economic and technological decentralisation. However, equally, it is possible to imagine a basically unchanged society using technical fixes to try to avoid such changes.

The technical fix approach has its attractions, in that it may help to avoid fundamental changes. After all, while some people see radical lifestyle changes as desirable ethically and politically, others would see them as unfortunate impositions. But change of some sort seems inevitable. In the belief that it is vital to try to steer a conscious path towards an environmentally sustainable and socially equitable future, rather than having changes forced on us, it seems wise to try to map out what sort of society might be aimed for.

Utopians through the centuries have tried to do this, by producing blueprints for ideal societies. They can be very inspiring, but they tend to ignore the thorny transitional problem of how to reach the ideal state. In reality what seems to be needed is a process of social negotiation of both means and ends. There is no one obvious way forward: rather there is a host of often complex tactical and strategic issues to resolve and a host of possible ways forward.

Alternative trajectories

The alternative society dreamt of by the counter-culture in the early 1970s had, as its technological base, a set of alternative technologies – the classic windmill, solar collector and biogas generator used in a small self-sufficient rural community. This dream of rural retreat lives on, but has it much relevance to the bulk of the world's population? Some greens think so, for no other reason than that they believe that the mainstream society is likely to collapse and so the best place to be is 'not underneath'.

But even assuming mainstream society continues, some sort of social and technical decentralisation would seem to be a valid goal. Large-scale centrally integrated systems, once seen as modern and efficient, are often now viewed as being hopelessly inflexible, prone to monopolistic or bureaucratic control and environmentally undesirable.

Does this mean that small is always beautiful? Certainly it can be. But we must be wary of simple technological determinism: technology does not define society. After all a decentralised society existed in the Middle Ages, based on small-scale wind, water and biomass, but few people would suggest that feudalism was a progressive state, whatever the technology. Neither does society totally define technology: there is a host of possible ways forward technologically – a whole series of technological trajectories based on different mixes, types and patterns of technological activity. Each may involve a slightly different balance between social and environmental concerns, reflecting different political and cultural attitudes. This is why it is so difficult to prescribe blueprints.

Maintaining diversity is a good ecological principle. There is room for a variety of approaches to sustainability tailored to specific social, cultural and environmental contexts, but there are also overall limits to the range of social and technological choices – primarily environmental limits. We have explored some of them in previous chapters in technical terms, such as the natural energy limit; and we have outlined some general criteria for sustainable technology. But within this 'sustainable development' envelope, there is a range of social and technological possibilities.

A key task for the future is to try to reach agreements on the right mixes in each situation, locally, nationally and globally, while recognising that each of these levels of concern interacts to some degree. This will involve some trading off between short and long term concerns, tactical and strategic considerations, and local and global perspectives.

Small is beautiful?

In order to make these rather abstract generalisations more concrete, let us look at the issue of technological scale, since it provides an example of the problem of balancing out some of these conflicting local and global, tactical and strategic concerns. The attractions of small-scale technologies, as initially promoted by the alternative technology movement in the 1970s, seem clear enough. Small-scale systems avoid losses in distribution from centralised power plants, and they seem likely to have lower environmental impact than large systems. They can also, in principle, be built and run by individuals or small groups, and they can therefore perhaps foster local self-determination and democratic control.

However, it is not clear if these potential benefits always materialise in practice. For example, local control may not always lead to optimal environmental choices in wider national or global terms. Remember the debate over wind farms in the UK. Not-in-my-back-yard (NIMBY) reactions may dominate to resist some potentially beneficial technologies, while other technologies of arguably less merit may be welcomed for purely selfish local reasons. It is not always the case that local residents take care of their environment.

Second, small-scale technology is not always environmentally appropriate. The basic physics of wind turbines implies that small diameter machines are much less effective that larger machines: as you may recall there is a square law for power output, so doubling the blade diameter quadruples the power output. Small machines can perhaps be more easily located near human settlements without having an adverse impact, but using larger machines, especially if they are grouped in clusters, can generate sufficient power to make it possible to site them remotely and send the power back to users on power grids. Remote location makes it possible to get access to high wind speed upland sites and, as you may remember, the power in the wind is the cube of the wind speed. So bigger machines in windy sites are very much more effective than the same total capacity of smaller machines at less windy locations.

This is reflected directly in the cost of generating power from small devices. In the mid-1990s a typical 1 kW domestic scale wind turbine costs around £3,000. Typical commercial scale wind farm machines of say 300 kW cost around £600 per installed kW (i.e. five times less per kW). The extra cost reflects the fact that small machines use more materials per kW installed and per kWh produced than big machines – more steel for the towers, copper for the generator, and so on. A comparative study of the 'energy balance' (i.e. input energy versus output energy) of a range of energy generation technologies compared the amount of energy needed to manufacture them (and their constituent materials) with the amount of energy produced in operation. It indicated that small 'stand alone' wind turbines, using battery storage, came out as the worst option of all (Mortimer 1991). So people who choose to run small wind turbines to try to attain some degree of self-sufficiency are not only paying a cash premium, but also imposing an extra environmental resource cost on the planet.

There is a further problem in terms of storage. The individual 'offgrid' wind turbine, run independently to supply power to a remote dwelling, will need some form of power storage since the wind does not blow all the time. Batteries are widely used, but they contain lead. Do we really want to bequeath our children yet more lead to deal with? By contrast, with a national network of grid-linked wind farms power can be shunted around the country, with each wind farm supplying power to the grid when it can. The grid can thus act as a form of buffer or store to compensate for local variations in wind power availability. This should more than offset any power losses due to transmission along the grid.

There are situations where local independent 'stand alone' generation is reasonable and indeed sensible – in 'off grid' locations where the alternative is to use diesel or petrol to fuel generators. There is a lot of interest in this level of activity in remote areas. However, there is also increasing enthusiasm for independent power among people who want to be self-sufficient, even though they might be able to use grid power. In the USA in particular there has been something of a rebirth of the 1970s alternative technology movement, with the emphasis being on 'getting off the grid' via self-help initiatives (Jeffrey 1995). This option is somewhat easier these days due to technological developments; for example, some 'home power' enthusiasts are using photovoltiac cells and other 'high tech' devices (Allen and Todd 1995).

The more traditional approach, as pioneered by the alternative technology movement, involved putting more emphasis on 'low technology', for example, making use of scrap materials to construct simple wind devices on a local do-it-yourself (DIY) basis, and thus avoiding the use of scarce resources. Certainly one of the still valid principles of the old alternative technology movement is that it is environmentally helpful to try to meet local needs from local resources. However, there are limits. It is probably much easier in terms of, say, biofuels like wood than for more complex engineering-based generating technologies like wind turbines, unless you are prepared to use relatively simple but perhaps less efficient homemade devices. There is clearly a role for local level DIY initiatives, but overall it seems likely that, in terms of small-scale local usage of renewables, in most cases the future will lie with professionally designed and built high performance machines and systems.

Off the grid?

Assuming that most of the machines used will be professionally designed, this still leaves the question of whether they can and should be run independently, that is off the grid. The analysis of scale factors above suggests that totally 'off grid' generation, for example, using small wind turbines, may be environmentally suboptimal as well as expensive, with the implication that it would be better to share common services via wind turbines linked to the grid so as to trade between areas and even out local variations in wind availability.

This conclusion may be changed to some extent by technological breakthroughs. In terms of cost, mass production of wind turbines is likely to reduce outlay and, traditionally, cost reductions are greatest when dealing with larger numbers of identical mass-produced products. So smaller machines may get cheaper faster than large turbines. Improved designs may also use less material resources. The development of new energy storage techniques and a move towards the use of hydrogen as a new fuel might also make reliance on local energy sources more credible.

Organisational changes might also improve the situation for smaller scale technology. In the UK, for example, following the liberalisation of the energy distribution network, it is now possible for consumers requiring more than 100 kW to contract with any supplier, using the grid as a 'common carrier'. Previously the only option was to buy power from the regional electricity company. Under the new arrangement groups of local consumers could in principle join together to bulk buy power from a 'green' energy supplier located anywhere in the country. The UK-based Body Shop cosmetics company has already pioneered a version of this idea: it has invested in a wind farm in Wales in an attempt to ensure that it puts back into the grid as much green energy as it uses from conventional sources.

This type of approach clearly represents a compromise between total self-sufficiency and national level integration via the grid. In principle you can get the best of both, with a distributed network of small projects being linked up by the grid, feeding power to local users and also to consumers around the country. On the consumers' side, the 100 kW minimum limit currently in force in the UK means that they can benefit directly only if they group together. But there are plans for this limit to be removed, from 1998 onwards, so that individual households could then contract with suppliers of their choice. On the power generation side, rather than having totally independent machines, many individual small wind turbines in a region, country, or even more widely could be linked up to export any excess power to the national grid. This would result in an integrated yet very decentralised network of power generation – local generation contributing to collective national power availability and collective provisions being available for local use. Local generators could well proliferate in this situation. Indeed, as in Denmark, some local consumers might decide to set up themselves to supply their own needs and sell any excess power via the grid, taking in power when the level of local generation was not sufficient.

Providing the necessary cabling and control systems for a decentralised system such as this would perhaps be expensive: one of the obstacles to the proposed lifting of the 100 kW limit in the UK is that it would cost an estimated £150–200 million to install the necessary national computer control system to handle (and charge for) the energy transactions. It might also be argued that such a system would not be very cost effective, given the small amounts of power coming in at each point in the network and the losses in transmission along the grid. As one critic put it, you might end up only warming the wires. But under most conditions the bulk of the power would actually be used locally, thus lessening this problem. Indeed it has been argued that meeting local needs by local generation from renewables should actually be given credit since it avoids the power losses associated with distributing electricity from centralised power plants to remote users.

The concept is known as 'embedded generation'. Local generators are located at the far end of the grid where grids are weak and where the power they produce is more valuable than that which has to be sent long distances (Mitchell 1995). Obviously the benefits of using 'locally embedded' generation will only accrue when local sources are producing a part of the net national energy requirements, i.e. mainly just what is needed locally. Once power has to be shifted around the country, either into or out of localities, this advantage is lost.

Local generation

Even so, the concept of local generation integrated into a national system is an attractive one, especially since it does not need to apply only to wind turbines. Local excess power from biomass fuelled plants, small hydroelectric turbines in rivers and streams, or from photovoltaic solar cell arrays on local dwellings and commercial premises could also be included. Micro hydro is of course a site-specific option, available only in some locations, but energy crops and photovoltaics seem likely to lend themselves to widespread use on the basis of local level decentralised operation, integrated with the grid.

Some local energy needs could also perhaps be met without requiring high quality and totally 'firm' grid supplies. For example, it seems foolish to use expensively generated frequency and voltage stabilised grid electricity just for heating. In reality, only a relatively few domestic and commercial devices actually need 50 cycle ac mains (most electronic devices convert it internally to low voltage dc) and some devices like freezers can cope with periods of disconnection. It could be that separate local grids will be developed for some uses, perhaps using lower voltage direct current generated from local renewables, topped up when necessary with power from the national grid.

As can be seen there is a range of electricity generating renewable energy technologies which could be used effectively at local level on a relatively small scale, as long as care is taken to choose the scale carefully and to try to integrate the system wherever possible. It is also worth remembering that electricity is only needed for some applications: the bulk of local energy needs are for heating, and there is a range of heat production options at local level.

However, there are some renewable energy technologies that can only be used on a relatively large scale – tidal barrages being an obvious example. The barrage at one time proposed on the Severn estuary in the UK would have had an 8 GW capacity. Offshore (i.e. deep sea) wave energy generation would be on a similarly large scale. Obviously there are also possibilities for smaller barrages and smaller onshore or inshore wave energy devices, but it does seem that any viable energy system using renewable energy would have to have a mix of large-scale and small-scale renewable energy technologies. Very roughly speaking, and depending very much on the pattern of society being considered, it might be possible to generate about as much from small-scale local technologies as from large centralised systems. But the precise balance is a matter for debate and negotiation – on the basis of technical, social, economic and environmental concerns.

Energy crops

The need for social negotiation in relation to technical, economic and environmental factors can perhaps be made clearer by looking at a further example of a new form of renewable energy – energy crops. As we noted earlier, the use of biomass as a fuel looks very promising around the world. The specific options vary from region to region: sugar cane has been used to produce alcohol in Brazil for use in cars, while in Europe oil seed rape has been used to produce so called 'green diesel'. The provision of transport fuel has been a high priority, but energy crops can also be used as a fuel for direct heat production (e.g. straw is used in many Scandinavian countries to power district heating networks) or for use in electric power generation (eg via multifuel boilers or gas turbines). In the UK there is much interest in the regular coppicing of quick-growing willow and poplar to provide wood chips for use in electricity generation via gas turbines (Macpherson 1995).

The use of crops for fuel raises a number of environmental issues. We mentioned some of them in Chapter 6. Perhaps the key strategic question is: should we really be using agricultural land for energy production rather than food production? Part of the reason this has happened in Europe is that, due to the use of high productivity intensive farming methods, there is a surplus in food production. As part of the EU's Common Agricultural Policy, some land has been 'set aside' to protect food markets. Using set aside land for energy crops would appear to be a better option than leaving it fallow, and it seems better to pay subsidies to farmers for growing something rather than for doing nothing. Energy crops are also seen as renewable energy sources – assuming that the rate of replanting equals the rate at which the crops are burned, then there is no net increase in carbon dioxide production, since plants take in carbon dioxide while growing. However, not everyone is happy with the Common Agricultural Policy. At the most general level, there are those who feel that it is obscene to 'set aside' agricultural land to protect profits when so many in the world are starving, although this of course begs the question of distribution.

Leaving this issue aside, some environmentalists argue that it would be better to switch to less intense 'organic' forms of farming, which would use the set aside land in a more environmentally sound way (Safe Alliance 1992). Not only would soil quality and produce be improved, but more habitats would be available for wildlife and there might also be a net energy gain, since the energy used in intensive farming may be more than would be generated by using the set aside land for energy crops. The objectors also fear that energy crops will become big business, so that there will be giant energy crop plantation dominating the landscape. They are also concerned about the impact of pesticides and fertilisers – although the proponents of energy crops argue that the use of these will be less than with conventional agriculture. Finally, there are concerns about possible toxic emissions from combustion. Greenpeace, for example, has mounted a 'Ban the Burn' campaign, focused mainly on the use of domestic and commercial rubbish as a fuel for power stations, but it could also be extended to the use of energy crops (Greenpeace 1992).

For our purposes, what is interesting about this debate is the balancing of interests among the four categories outlined in our model of interactions. Farmers are keen on energy crops as a source of revenue and employment; investors see them as potentially profitable; and some consumers are attracted to the idea of 'green fuel' for cars or 'green electricity' from plants. However, although some environmentalists support energy crops as a useful new renewable source, others are concerned about local impacts and the land use implications, with potential impacts on water resources also being seen as a problem in some areas.

These are not trivial issues: energy crops could become the largest single renewable energy resource. For example, as we noted earlier, a study of the UK resource by ETSU suggested that, if fully developed, energy crops like short rotation coppicing of willow and polar could eventually provide 150 TWh p.a., or around half of the UK's electricity requirements (Department of Trade and Industry 1994b).

Whereas the debate on wind farms has now been underway some while, that on energy crops is really only just beginning. Assuming that some use of energy crops is accepted, then the questions are: how much, in what way, where, and by whom? There would seem to be a need to negotiate sensible trade-offs and compromises between the various technical and environmental factors. Local impacts will have to be carefully thought through and discussed, leading to modified planning rules to avoid poor siting and overlarge projects, and to improved emission controls.

Beyond that there would seem to be a need to discuss the overall contribution of energy crops to world energy production. Some energy scenarios, like the global renewable energy scenario produced by Johannson *et al.* for the 1992 Earth Summit, rely on them heavily; others, like the fossil free energy report produced for Greenpeace, are more wary, with the potentially negative impact on biodiversity being seen as a key issue (Johansson *et al.* 1993; Greenpeace 1993).

Structuring the debate

Local and global trade-offs between technical, social and environmental concerns of the type discussed above are often at present made primarily in economic terms, on the basis of cost. Attempts have been made to add environmental costs into the equation more effectively, for example, by the use of cost benefit analysis, with financial values assigned to each element. However, this is difficult. How can you value a landscape or a tree? 'Contingent valuation' techniques, in which people are asked to express their individual estimates of comparative value (eg how much they would be willing to pay to retain some amenity) are inevitably limited, although they do at least begin to open up some form of social negotiation process.

The economic and industrial forces that shape decision making about technology are very powerful and it is hard to see how they can be easily counterposed unless there are equally powerful influences at work. In principle environmental planning legislation and regulatory agencies such as the Environmental Protection Agency in the USA and the UK's Town and Country Planning system provide a means for allowing divergent views some hearing, for example, via local public planning inquiries or public hearings on major projects. However, these usually operate within the context of already established higher level technology policies. Thus public inquiries over the desirability of a motorway or a power plant will not normally be allowed to discuss wider issues of transport policy or energy policy, only whether the specific project is acceptable. Even so, environmental concerns are gradually beginning to influence the way in which governments define their technological and industrial policies and companies are also beginning to take these issues on board, if for no other reason than to try to anticipate impending government regulations and legislation.

The purpose of this book is not to explore the details of such developments or to delve into the rapidly expanding field of environmental economics: this would require a whole book on its own. However, it does seem that the sort of criteria for sustainable technology outlined in Chapter 3 are beginning to be put on the corporate and governmental agenda.

In part this has been possible because rather than imposing costs on the various human actors in the game, a switch to sustainable approaches may in fact also offer opportunities – for example, opening up new markets for green products, providing employment options for those made redundant from dying traditional industries, and generally creating wealth. Thus, Sir Crispin Tickell, Convenor of the UK government's Panel on Sustainable Development, has argued that 'People have become rich making a mess over the last two hundred years. In my judgement they could become even richer still clearing it up over the next two hundred' (Tickell 1994: 33).

Not everyone accepts this optimistic interpretation. It is clear that adopting a sustainable approach to technology, and to economic development generally, could create many new jobs and underpin a more stable world economy. However, as we have seen, overall there may be a need for each of the human groups to accept some burden in order to achieve environmental sustainability.

Conclusion

Technology offers both threats and promises. It can ruin the planet, but possibly it can also save it. However, technology is really just a dead assembly of things organised and used as a tool for some human purpose. There is certainly a need for careful and concerted assessment in that some technologies may be considered to be socially and environmentally inappropriate, but only the most extreme critic would say let us do away with all technology. The real need is to be selective and try to develop a more balanced and sustainable pattern of living, within the constraints imposed by the ecosystem.

This balance is not going to be easy to achieve. Even if people are so minded, it is difficult to know precisely how to defend the environment. After all, although interconnected, the natural world is not just one integrated system. There are thousands of species which may need protecting, thousands of ecological problems that need attention. How can environmentalists choose which to focus on? Must they play god by selecting to defend one species instead of another? Is there not a risk that human intervention may become partial, biased and counter-productive?

Some hardline environmentalists have come out against major renewable energy projects like tidal barrages, on the argument that these would seriously disturb wildlife. The debate is a complex one: while some species might suffer a degraded habitat, others might actually benefit from the local ecological changes. The balance is hard to judge. It could be that other technologies would be a better choice, but there is a need for a full analysis of the options and their impacts. In this situation, the adoption of an overly doctrinaire approach to defending specific species and habitats might be unhelpful.

It must also be the case that some environmental problems are likely to be more serious than others in the long term. Given that there are limited resources of money and expertise, there has to be some degree of focus. But how can assessments be made of which problems are going to be the key ones? The standard argument is that the focus should primarily be on any environmental impacts which seem likely to be irreversible. However, this may not always be easy to identify. What may seem a trivial issue could turn out to be central. In this situation much reliance must inevitably be placed on what scientific research can impart. Unfortunately, however, science can only go so far. In the end it will inevitably come down to subjective human judgements.

Overall, what seems to be needed is a comprehensive approach to environmental protection that seeks to balance all interests and maintain overall biodiversity – and this is not easy. In previous chapters some of the ultimate environmental and resource limits were surveyed and an attempt was made to develop some general strategic principles and criteria for sustainable technology. But this only provides an outline guide for specific choices. A process of social negotiation is needed to make such choices with the natural environment having, as it were, a proxy vote.

Some might say that in a world driven by private and even collective greed it will be difficult to bring about positive improvements. At present, technological and industrial development is being driven at ever-increasing rates by the competitive world economy, fuelled by the profit motive and coupled with people's expectations for more and more consumer goods and services. Some would say that the technology or the rate of change is the central problem, while others would argue, more fundamentally, that it is the underlying social motivations, and the economic and industrial structures that have been created as a result, which are the basic cause (Braun 1995).

There is no way that any objective conclusions on such issues can be reached in this book: it is a matter of viewpoint. Clearly technology can reflect both the best and the worst human motivations. This is not to suggest that technology is a fixed set of items which can either be used or abused. The way technology is developed and deployed and the choice of technology are a social and political processes, reflecting the conflicting interests and power of rival groups in society. This has always been the case. What has changed in recent years is that, to return to the model of interaction outlined in Chapter 1, a new factor has become crucial – the natural environment. If humans are to live successfully on this planet it becomes increasingly vital that their technological choices take account of more than just the partisan interests of the conflicting groups or the interests of power elites, and include the fundamental environmental limits and constraints. In practice what seems inevitable is that a process of prioritisation of the issues and options will have to emerge. What is not clear is the political issue of who will do this, and on what basis.

Summary points

- There are many paths to sustainability, with a range of mixes of technology.
- Technology can serve a variety of different patterns and scales of social organisation, although there may be environmental limits on what is viable.
- Choosing the best options for any particular society ought to involve a process of social negotiation, with environmental interests playing a key role.
- Environmental issues are becoming increasingly important, but it is often hard to know what are the key issues. There are many uncertainties and conflicts facing mankind should it choose to try to move towards a sustainable future.

Further reading

There is a vast and growing literature on sustainable development and the associated economic, industrial and technological issues. For a useful introductory review of some of the key issues see Sharon Beder, *The Nature of Sustainable Development* (1993, Scribe, Australia).

Some of the general economic and development issues are discussed in the series of books and reports produced by the New Economics Foundation, London, such as Paul Ekins' introductory *Atlas of New Economics* (1992, Hutchinson, London).

Energy economics is a rapidly expanding field. The Blueprint series of studies from CESERGE, the UK-based Centre for Social and Economic Research in the Global Environment, produced by Professor David Pearce *et al.*, provides a useful introduction.

A critique of some aspects of the conventional approach to energy economics can be found in Andrew Starling's paper 'Regulating the electricity supply industry by valuing environmental effects', *Futures*, 24, (10): 1024–47, 1993.

16 Conclusions: the way ahead?

- **Resolving conflicts by negotiation**
- **Bottom-up initiatives**
- **Criteria for sustainability**
- **Human responsibilities**

The way ahead for any attempt to attain sustainability must involve action at all levels, locally and globally: 'thinking globally and acting locally'. Drawing on the model of interests introduced at the start of the book, this chapter uses some of the material from the case studies and examples in the rest of the book in order to assess some of the opportunities that exist for negotiating trade-offs between conflicting human interest and local and global concerns. If sustainability is to be achieved this sort of negotiation process seems vital, the final issue being whether human beings are actually capable of negotiating ways to solve the environmental problems which they have created. If not, the future could be bleak.

Negotiating the future

It is not the aim of this book to provide a blueprint for what specifically should or could be done to try to solve the environmental problems that face the world. Various specific technological options have been reviewed and it may be felt that some might add up to a viable sustainable energy package. However, the main point to emerge is that there is a need to choose among them and to negotiate exactly how they are to be used, as part of a wider process of moving towards a sustainable future. With this process of negotiation in mind, some of the general themes and issues from earlier chapters can be usefully brought together so as to indicate some possible starting-points and possible ways in which conflicts of interest might be overcome.

The model of conflicting interests introduced in Chapter 1 highlighted the need for a process of negotiating between the various human interest groups (producers, consumers, shareholders) and the interests of the planet, with the latter being given priority. Thus in Chapter 3, where we developed a list of criteria for sustainable technology, it was suggested that human sensitivities in relation to the environmental impacts of energy projects were only part of the issue. If human beings need power, then it should be human beings that paid the price. Thus, the environmental costs which mattered most should not necessarily be those relating to human perceptions of the natural environmental (e.g. visual intrusion in scenic landscapes) but those which might affect the rest of the ecosystem.

However, there are degrees of impact: what must really be avoided are those which produce irreversible adverse changes. But in many cases it is hard to decide what the long-term effect might be of what would seem to be small-scale local impacts, especially when combined, possibly in unexpected ways, with other small impacts. Concepts like biodiversity may help to keep some sort of balance, but in the end the precautionary principle seems important: it suggests that any potentially dangerous changes should be avoided if the likely outcomes are unknown.

While overall priority can perhaps be given to the environment, how can priorities among the various human groups be arranged? Energy projects are likely to provide benefits to some but costs to others. In Chapter 12 we looked at the issue of wind farms. The environmental benefits are clear and they accrue, in general, to everyone in terms of reduced air pollution and avoidance of global environmental problems. But the economic benefits accrue to developers and their investors, while the environmental/amenity costs fall primarily on local residents. It was suggested that one way to resolve this imbalance would be to encourage local ownership of wind projects.

Unfortunately, there are not always potential solutions available; in some cases all the human groups may have to accept real costs in order to defend the planet. In Chapter 13 we discussed whether it would be possible to continue with economic growth indefinitely and came to the conclusion that, although it might be possible technically for the foreseeable future, there would probably also be a need for social changes with, at the very least, less emphasis on the quantity of material consumption and more on the quality. However, by way of compensation, it was noted that a switch to a more sustainable approach might help resolve the problem of unemployment – at least in some sectors of the economy.

Of course it would not always be easy to make these changes. In Chapter 3 we looked briefly at cars and pointed out that, while in the short term consumers and producers had a common interest in continuing with existing types of vehicles, in the longer term this would not benefit either group. Everyone has to breathe and the adoption of a more sustainable approach seemed inevitable at some point.

What about the other main human interest group identified in our model – the shareholders – i.e. those who benefit from investing in production using natural resources? Will profits and dividends be affected? The simplest reply is that existing patterns of commerce are environmentally unsustainable so there is no choice, in the long term, but to make changes. Investors will have to accept their share of the burden. More subtly, it is sometimes argued that a sustainable economy could be as profitable, or nearly, as the existing one. But it would be different – with more emphasis on what has been called 'ethical investment'.

Currently there are many investment agencies offering shares in projects that have been subject to assessment on environmental grounds. For example, in the UK the Wind Fund is offering shares in selected wind farm projects, with a minimum stake of £300. Backed by a Danish bank, the aim is to widen the base of shareholding and to stimulate wind farm development. The local community-based wind co-ops in Holland and Denmark have similar intentions. Local environmentally oriented investment of this sort seems to be spreading elsewhere: 20 per cent of the 1.2 GW of wind turbine capacity established in Germany by 1996 was owned by local community groups.

In part these trends reflect a dissatisfaction with the traditional form of investment – whether public or private. Nationalised, state-owned companies are seen as hopelessly bureaucratic, centralised and inflexible, while the large private corporate monopolies are viewed as unaccountable, insensitive and greedy. There is much that can be done to remedy this situation by way of institution innovation, increased accountability and so on at the national governmental and corporate level, but equally there is a role for local level initiatives.

Global solutions, local initiatives

While difficult trade-offs and bargaining between human and environmental concerns will be required at all levels, it seems clear that painful negotiations can perhaps be more effective, and possibly more equitable, if carried out at the local level. In part, this is what has been enshrined in the Local Agenda 21 programme initiated following the UN Earth Summit in 1992 (United Nations 1992).

Around the world many national and local environmental organisations have become active in this programme, including city councils and local government bodies. There are many independent local initiatives, projects and campaigns, many of which are developing practical sustainable development projects at grassroots level. Some of the organisations listed in Appendix II can provide up-to-date details of what is inevitably a rapidly changing scene. But there are already many projects in the sustainable energy field around the world – East and West, North and South. For example, it is heartening to see, in the coverage provided by INforSE's journal *Sustainable Energy News*, grassroots renewable energy projects in Eastern and Central Europe.

It is also good to see local projects springing up in economically depressed areas in the West. For example in the UK, there are projects underway that are designed to revitalise areas which were once reliant on coalmining, with the use of local renewable energy sources playing a key role. Earth Balance in Northumberland is making use of wind and biomass and hoping to create employment opportunities in an area once dominated by coalmining, while Broughton Energy Village, on an old coalmine site in the Nottingham area, is using wood chips from the nearby Sherwood Forest. The Earth Centre, which is being developed on an old coalmine site near Doncaster in South Yorkshire, is a 120-hectare demonstration of 'sustainable living' with plans to make use of local biomass as a fuel. It has benefited from support from the UK's Millennium fund.

As the director of the Earth Centre project put it, 'If sustainable development is to work, it has to start with areas like this which suffer from high unemployment (10,500 jobs have been lost in the last 10 years) and which have experienced the worst excesses of many traditional, polluting industries' (*RENEW* 1996).

As can be seen from these examples, the motto 'thinking globally and acting locally' has been taken seriously. There is much that can be done at the local level. Of course some of the negotiations concerning global concerns can only be made at international level, but even here a local input can be made. For example, over the past few years a series of basic environmental criteria have been developed via a consensus process at grassroots level in Sweden, under the name of the Natural Step (Greyson 1995).

The 'system criteria' that have emerged (see Box 6) emphasise the need for a sustainable approach in all sectors, and they have been adopted by several major Swedish companies. As can be seen there are some links with the criteria developed earlier in this book, although the Natural Step criteria are more fundamental. The Natural Step concept is now being internationalised – by inviting other organisations around the world to promote and try to win acceptance for these basic principles. The UK version is being promoted under the name of the Planet Pledge. Clearly, it is a very general set of criteria, but it begins to establish a new agenda in a 'bottom-up' way.

Box 6

Natural Step: system criteria

1 Substances from the Earth's crust must not systematically increase in nature. (This requires a considerable reduction in our dependence on mining and the use of fossil fuels.)

2 Substances produced by society must not systematically increase in nature. (This requires phasing out of persistent unnatural substances such as CFCs and PCBs. It also requires reduction of the production of naturally occurring substances such as carbon dioxide and sulphur dioxide.)

3 The productivity and diversity of nature must not be systematically diminished. (This requires sweeping changes in our use of productive land, for example, in agriculture, forestry, fishing and the planning of societies.)

4 Humanity must achieve the just and efficient use of resources in society. (This requires basic human needs to be met with the most resource efficient methods possible, including a just income distribution. Do more with less.)

Natural Step is only one of several attempts to develop and promote basic criteria for sustainability. Others are less general and aimed at specific target groups such as engineers, designers and other professionals involved with technological development. For example, in *Greening Business* (1991) John Davis outlines a set of criteria for engineers (see Box 7).

The advent of the concept of 'green product design' has led to a similar set of criteria emerging for designers (Blair 1992), while Mike Cooley (1987), an exponent of the idea of 'socially useful production', has produced a set of criteria which emphasise the need to consider not only products but also the production process and the role of producers within it (see Box 8).

Box 7

The challenge of sustainable development

Outline code of engineering for sustainable development

- The primary purpose of civil application of engineering is to harness the forces and resources of nature for the benefit of mankind and the environment.
- Benefits should be as widely available as possible, so:

  1 Natural resources should be used as efficiently as possible and renewables are preferred.

  2 The financial cost of users, over the whole life of the product, should be as low as possible.

  3 The product should not demand exceptional user skill.

  4 Production and use should not dehumanise people.

- The highest technically achievable standards and energy efficiency should be aimed at.
- In evaluating the benefits/costs, long term mass application effects must be considered as well as short term, limited application effects.
- Every undertaking must respect human rights and human dignity. Engineering should not be carried out with the intention of advantaging some people at the expense of others; it should as far as possible advantage the disadvantaged.
- A 'total systems approach' should be adopted: maintenance, repair and reconditioning should be facilitated, and materials should be recyclable.
- Measures should be taken to prevent misuse wherever possible.
- Ultimate disposal of a product must be considered at the design stage, and plans for acceptable solutions prepared.
- Knowledge regarding safety to people and the natural environment should be freely shared.
- No relevant information regarding use/application should be withheld.
- Respect must be paid to all patented inventions and registered designs.
- Involvement in the design, development or production of illegal goods is forbidden.

Source: John Davis. 1991. *Greening Business–Managing Sustainable Development*, Basil Blackwell, Oxford.

Criteria like these can play a useful role in 'raising consciousness' among practitioners. At the same time there remains a need for all the usual forms of advocacy, campaigning and lobbying to help win wider social and political acceptance of the new paradigm of sustainable technological and social development which is emerging around the world.

Box 8

Socially useful production

A tentative list of those attributes, characteristics and criteria which constitute socially useful production. It is not suggested that all these will be present in any particular socially useful product or production programme, but rather that some of these are key elements within it.

1. The process by which the product is identified and designed is itself an important part of the total process.
2. The means by which it is produced, used and repaired should be non-alienating.
3. The nature of the product should be such as to render it as visible and understandable as is possible and compatible with its performance requirements.
4. The product should be designed in such a way as to make it repairable.
5. The process of manufacture, use and repair should be such as to conserve energy and materials.
6. The manufacturing process, the manner in which the product is used and the form of its repair and final disposal should be ecologically desirable and sustainable.
7. Products should be considered for their long-term characteristics rather than short-term ones.
8. The nature of the products and their means of production should be such as to help and liberate human beings rather than constrain, control and physically or mentally damage them.
9. The production should assist co-operation between people as producers and consumers, and between nation states, rather than induce primitive competition.
10. Simple, safe, robust design should be regarded as a virtue rather than complex 'brittle' systems.
11. The product and processes should be such that they can be controlled by human beings rather than the reverse.
12. The product and processes should be regarded as important more in respect of their use value than their exchange value.
13. The products should be such as to assist minorities, disadvantaged groups and those materially and otherwise deprived.
14. Products for the Third World which provide for mutually non-exploitative relationships with the developed countries are to be advocated.

15 Products and process should be regarded as part of culture, and as such meet the cultural, historical and other requirements of those who will build and use them.

16 In the manufacture of products, and in their use and repair, one should be concerned not merely with production, but with the reproduction of knowledge and competence.

Source: Mike Cooley, 1987, *Architect or Bee?*, Chatto & Windus, London.

Views inevitably differ as to what represents the best way forward for individuals and organisations. While some focus on national and international level issues via pressure groups and other forms of lobbying, others see local 'self-help' initiatives and community-based campaigns as the best route: acting locally while thinking globally. The individual's level of commitment can vary from improving personal lifestyles or working with others to set up local community-based energy projects through to changing the way in which employers or other agencies operate and campaigning on international environmental issues. (See Appendix II for a list of key contact points for further personal involvement.)

In the final analysis, most people in the 'green' movement believe that sustainability can only be achieved if widespread agreement can be reached on what needs to be done: quite apart from being inequitable and undemocratic, draconian solutions imposed from above by elites are likely to create more problems than they solve. Grassroots initiatives may at times look weak by comparison with the scale of the problems and the power of the large organisations that currently shape world affairs. But, as the contacts list (Appendix II) indicates, networks are growing up around the world, often making use of Internet computer links, that may offer at least some solutions to the problem of thinking globally and acting locally.

New ideas?

In a presentation to a conference on alternative technology in 1994, Peter Harper, one the early pioneers of the concept, concluded as follows:

> I don't think the problem of sustainability is best solved by better technology. It probably can be – and probably will be. But this is a tiresome, bloodless, round-the-houses-route when really positive and culturally deft solutions are right in front of us.
>
> (Harper 1994: 18)

He was referring to radical changes in lifestyle. He was well aware that this might threaten some people and felt that, before becoming more widely acceptable, the new types of living would probably have to be pioneered on the fringes of society, for example, in the various alternative communities that have grown up around the world. He saw the experimental residential communities like the Centre for Alternative Technology in Wales, Findhorn in Scotland and The Farm in Tennessee as 'nurseries' for developing new ways of living. At the same time, in some cases these centres could also provide demonstrations of how alternative technologies could support the new sustainable lifestyles. The Centre for Alternative Technology, for example, is a major demonstration and exhibition site attracting over 100,000 visitors each year, but it generates most of its power from local sources – chiefly water, wind and solar power.

Not everyone will be prepared to go to such extremes. For most of us, life involves relatively conventional careers and communities. But even here changes are occurring; contemporary economic pressures seem to be forcing changes in family and community living patterns.

Perhaps most important are changes in ideas. For sustainability to mean more than just a desperate series of technical fixes, mankind seems likely to have to take on board a new view of the global ecosystem as an entity of which human beings are a part. In effect we need a new 'paradigm' – a new framework for thinking about ourselves and our relationships to the rest of the ecosystem. As Capra has suggested, we may have reached a key turning point in human development (Capra 1982).

The 1990s have seen a whole range of new ideas and world visions emerge, many of them sharing an underlying holistic ecosystem view. However, there is often a lack of coherence, a degree of naivety and a potential for conflicts. Certainly an eclectic mixture of ideas and motivations has emerged. A symbolic rejection of materialism by the affluent jostles with economic protectionism. Romantic utopianism is sometimes coupled with thinly disguised elitism. Libertarian sentiments compete with a belief in the need to reimpose 'natural order'.

Even so, it seem clear that with the demise of some of the older ideologies that shaped the twentieth century and with the new millennium approaching, a new awareness is beginning to dawn, if only, as yet, on the fringes. Ecologists are developing new ideas about biodiversity and bioregionalism and our scientific understanding of ecosystems is gradually improving. In parallel, and more practically, grassroots initiatives are throwing up a new awareness of how local communities can respond to environmental problems and new ideas are emerging from environmentalists involved with campaigns to protect endangered species and habitats. At the same time, a new appreciation of the knowledge and wisdom of native peoples is emerging, along with new philosophies concerning the relationship between mind, body and spirit.

The challenge for humanity

Perhaps the central issue for the future is whether a viable and widely acceptable new awareness can emerge from this complex social, cultural and in the end political process of ideological renewal and, if so, whether mankind can act on it effectively in time to avoid disaster.

Some people fear that human beings are unlikely to change their ideas in time, in which case the prospects for the future may be grim. However, some technological optimists believe that technology may enable changes to occur quickly. Some even believe that it might be possible genetically to re-engineer people to improve human mental capacities: a technical fix approach of cosmic proportions.

At the other extreme, some ask whether there is a need to change so as to be able to take responsibility for planetary survival. The middle ground position is that mankind can and should take on a stewardship role in relation to managing the global ecosystem, since human beings have disturbed its natural functions to such an extent that it cannot sustain itself without our conscious intervention. However, some New Age thinkers see this as hopelessly arrogant and believe that mankind should abandon all pretence of being able to 'control' the situation in which it finds itself and leave everything to be resolved by the self-regulating global ecosystem, in the form of Gaia. On this view 'nature knows best', although whether the end result will be beneficial or not for human beings is less clear: as the major irritant in the system it might be that human activities will be constrained.

Machines to the rescue?

Some New Age thinkers even seem to believe, as Kevin Kelly has suggested in his influential book *Out of Control: the New Biology of Machines*, that the future will be resolved not just by natural systems alone but by a process of co-evolution involving the added human innovation of integrated cybernetic 'intelligent computer' based networks. Kelly is adamant that this is not a proposal for letting machines 'take over'. For him, the world of the 'born' (i.e. living beings) and the world of the 'made' (i.e. intelligent computers and machines) will co-exist and create a new synthesis which will be able to ensure global ecosystem survival. But neither would be in control (Kelly 1995).

Kelly's idea in effect brings us full circle: if the main problem in the relationship of human beings with the rest of the ecosystem is the way our technology is used, then, for him, the answer is to let technology operate in different ways. Human beings should not use it to try to control nature, as they have attempted in the past, but should let it help to resolve the problems which they have created.

In essence, Kelly sees machines like intelligent computer systems, freed of human control, as providing the missing element in the contemporary ecosystem mix, since, unlike human beings, they can operate more like the other elements of the ecosystem, that is unconsciously and, in effect, blindly. For him, the automatic, instinctive and relatively simple behaviour by the individual components or elements of complex systems, whether bee swarms, ant colonies or complete ecosystems, is the key to the survival of the system. In effect, what Kelly is arguing is that computers and integrated networks of intelligent machines can, if freed from human control, create a new layer of autonomous activity in the living self-regulating Gaian ecosystem. This has as its aim the survival of the overall ecosystem system, not the individual components, human or otherwise: the addition of the new manmade elements will simply improve the capacity of the overall system to respond to change.

In this sort of future, if it ever came about, partisan human aspirations would certainly be less of a dominant influence. But so too might the human capacity for imagination, vision, growth, co-operation and creativity, not to mention humour. In *Architect or Bee*, Mike Cooley (1987) argues that machines can never substitute for human tacit knowledge and vision, and he quotes Karl Marx as follows: 'What distinguishes the worst architect from the best of bees is namely this. The architect will construct in his imagination that which he will ultimately erect in reality' (Marx 1974: 174).

Not surprisingly, Kelly's ideas, with their emphasis on the merits of beehive or ant colony mentality, have met with strong resistance from a range of critics, including those from the political left (Barbrook 1995). This is hardly surprising given that it could be argued that what Kelly describes already exists to some extent in the form of the global capitalist economic and industrial system. Arguably, this is unconcerned about the fate of individuals or even individual countries: its only goal is the survival, self-replication and growth of capital. Whether the vast integrated network of technology and economies which we have established can be seen in any sense as compatible with the Gaian self-regulatory ecosystem is unclear. The global techno-economic system's concern for its survival might lead it to constrain ecologically dangerous activities, but equally it could just continue blindly to the destruction of the ecosystem.

Certainly, this seems a strong possibility, the implication being that human beings, who created this system, must therefore try to constrain, redirect or even dismantle it. On this view, far from being a resolution of the problems created by the human use of technology, Kelly's cybernetic prescription would seem to involve forcing technology to play an even more central role, with human beings finally abandoning any pretence of responsibility.

The alternative seems clear: if cybernetic visions of automatic self-regulation do not appeal, and the global techno-economic system that mankind has created cannot be left to go its own way, then mankind will have to learn to deal with the situation itself, consciously and co-operatively.

Technical fixes can play their part. Computers can help people model, analyse and manage complex systems and computer intelligence may become a useful tool. Indeed without computing power it would probably be impossible to cope with the vastness of the environmental problems that now exist (Young 1993). This said, surely in the end it is up to mankind to find the way to live on this planet without destroying it. Unless, that is, you accept the ultimate technical fix view and believe that at some point mankind will have the means, as a species, to abandon this planet entirely, and start all over again somewhere else. The colonisation of space may be a worthy longer term project, as some environmentalists have argued (Deudney 1982), but really, for the present, human beings have to try to resolve the problems they have created here and not run away from them.

For the moment at least, there really is 'only one earth', and mankind must learn how to live on it without destroying the planet or itself.

Summary points

- Longer term vision must replace short-term partisan concerns if environmental degradation is to be avoided.

- Balancing the interest of people and planet requires intervention at all levels – from the global to the local.
- Bottom-up initiatives may be able to provide some unique contributions and provide a context for experimentation with new lifestyles.
- New viewpoints on the relationship between people and planet and new criteria for the choice of technology need to be developed.
- In the end it is the responsibility of human beings to solve the problems they have created.

Further reading

There are many ways in which you can keep abreast of the sort of developments discussed above. For example, the US-based Worldwatch Institute's annual 'State of the World' reports are a good source of data and analysis on current overall developments; it also produces reports on specific topic areas. At the grassroots level there is a host of journals and newsletters on green issues and local initiatives; take a look around your local alternative bookstore. In the UK one of the more thoughtful environmental journals is *Real World*. SERA's *New Ground* is also an excellent source of news and analysis from a radical perspective. The contacts list in Appendix II provides some other starting points, including the various new Internet-based information services.

Glossary

acid rain Acidic rain produced as a result of the release into the atmosphere of acidic gases such as sulphur dioxide, generated by the combustion of fossil fuel in power stations and cars.

end use energy The energy actually consumed at the point of use.

global warming The possible increase in average global temperatures as a result of an enhanced 'greenhouse effect' due to the release of gases such as carbon dioxide and methane into the atmosphere: global warming is one element in the resultant process of 'climate change'.

Kondratiev cycles The 'long waves' (i.e. cyclic patterns) in global economic activity identified by Kondratiev, and interpreted by some subsequent economists as being due to regular bursts of technological innovation.

nuclear fission The process of splitting the nucleus of certain atoms (e.g. uranium) with the resultant release of heat and radiation, as in atomic bombs or nuclear reactors.

nuclear fusion The process of fusing together certain light elements (e.g. hydrogen) to yield heat and radiation, as in the H-bomb and the yet to be fully developed fusion reactor.

primary energy The energy in the basic fuels or energy sources used, e.g. the energy in the fuel fed into conventional power stations.

renewable energy Energy sources such as the winds, waves and tides which are naturally replenished and cannot be used up.

strategy A plan of action based on high level goals. Strategic considerations are longer term and more fundamental than 'tactical' consideration. The concept derives from military thinking: strategy is about the overall war aims and plan while tactics are about specific battles.

sustainable development Technological, economic and industrial development patterns which are environmentally and socially sustainable.

sustainable fix A more radical technical fix which may go further towards a more comprehensive and lasting solution, e.g. renewable energy.

technical fix A technical solution to a social, environmental or technical problem which tends to deal with symptoms rather than causes and often creates further problems elsewhere or at a later date.

Appendix I

Exploratory questions

To help you get to grips with some of the issues discussed in this book, here are some general questions which you might like to explore. Some pointers are included after each, but these are open-ended questions, designed to get you thinking about the wider issues in more detail. You may find that you will want to follow up some of the further readings mentioned at the end of each chapter.

1 *Is it really necessary to make major changes in the way in which energy is generated and used, or can minor adjustments suffice?*

There is a range of technical fixes on the energy use side, but it seems likely that the use of fossil fuels will be environmentally unsustainable and certainly in the longer term there will be a need for new energy sources, even if energy conservation is taken seriously. (Chapters 1, 2, and 4 provide some of the key arguments.)

2 *What are the technical options for sustainable energy supply technologies?*

Although not strictly sustainable in the longer term, nuclear power is seen by some as a possible candidate for non-fossil energy supply. But there remain problems with nuclear fission, including the waste storage issue, and fusion looks like a long shot, with its own problems. By contrast, the renewable energy sources look promising although they too have problems. (Chapter 3 sets out the basic criteria, Chapter 5 looks at nuclear power, Chapters 6 and 7 at renewables.)

3 *Can new sustainable energy technologies replace the existing range of energy technologies?*

Technically it seems credible for renewable energy, coupled with conservation, to meet human needs into the far future, if these options are developed quickly. However, there are powerful vested interests in the technological status quo, and fossil fuel prices are relatively low, making change difficult. There is also a range of other implementation problems and a need to win public acceptance. (Chapters 8 to 12 review the prospects for sustainable energy and look at some of the problems.)

4 *Sustainability will never be attained unless and until major social, economic and political changes have occurred in terms of redistribution of power and wealth. Do you agree with this statement? If not what are the counter arguments?*

This is a version of the radical view that political and economic power determines all else and that tinkering on the margins, for example, with new technologies, will be of little use since the current socio-economic system is fundamentally flawed.

A possible counter argument is that technology shapes society to some extent and if less damaging technologies can be fed into the system, it will change. Another argument is that the system will reform itself since otherwise it will be doomed. Whether this process would benefit all concerned is however unclear. (Some of these issues are discussed in Chapters 13 to 15.)

5 *How can the necessary changes be brought about?*

Assuming that you do not feel that the situation is hopeless or that only drastic action will suffice (see question 4), the possible responses range from 'letting the market identify viable new options' (if you subscribe to the free market viewpoint) through to 'grassroots campaigning' to change viewpoints and introduce new practices and priorities. In between is a range of options, including personal and professional involvement with the process of developing and deploying sustainable approaches. (Chapter 16 summarises some of the options.)

6 *Is there hope for the future? Can sustainability be achieved?*

Mankind is technically ingenious and has also been able to develop a range of different patterns of social organisation when faced with changed circumstances. The environmental problems that lie ahead look quite serious to many. If these people are right, it remains to be seen if the necessary technical and social changes can be made sufficiently quickly to avert environmental disaster.

In terms of technology, fossil fuels remain relatively cheap and they will be with us for many decades, whatever strategy is adopted. So there is much to do to ensure they are used more efficiently and cleanly. But in the years ahead pressure to phase them out is likely to grow as environmental problems mount. There will also be much to do in terms of developing the sustainable alternatives. (The contacts list in Appendix II provides some starting points if you wish to get involved.)

Appendix II

Contacts

There are many organisations around the world which are trying to support moves towards a sustainable energy future. International environmental pressure groups like **Greenpeace** and **Friends of the Earth** have campaigns to this end and the various national and international governmental agencies can provide information on specific policies and programmes. In the UK one of the best sources of information on sustainable energy technology is the **Energy Technology Support Unit** at the Harwell Laboratory, Didcot, OX11 0RA. World Wide Web http://www.etsu.com. The US **Department of Energy** offers similar services: see their renewable energy and energy efficiency service on the World Wide Web: http://www.eren.doe.gov/.

For independent analysis and up-to-date reports on developments you may find it useful to contact the following:

CREST, Centre for Renewable Energy and Sustainable Technology, 1200 18th St NW 900, Washington DC 20036, USA. Tel: (202) 289 5370. Fax: (202) 530 2202. e-mail: <info@crest.org>. World Wide Web pages: http://solstice.crest.org. CREST's WWW-based information service (SOLSTICE) is excellent and will link you in to most of the Web sites around the world. It also produces a newsletter which is available free via e-mail.

INforSE, the International Network for Sustainable Energy, PO Box 2059, DK-1013, Copenhagen K, Denmark. Tel: +45 33121307. Fax: 45 33121308. e-mail: <info@nn.apc.org>. INforSE produces a newsletter *Sustainable Energy News* which is also available electronically via CREST.

NATTA, the Network for Alternative Technology and Technology Assessment, c/o EERU, Open

University, Milton Keynes, MK7 6AA, UK. Tel: 01908 654638. Fax: 01908 653744. e-mail: <S.J.Dougan@open.ac.uk>.
NATTA produces a newsletter *RENEW* on technical and strategic developments in the renewable energy technology area, available on subscription. Parts of it are also available free as 'Renew On-Line' on the World Wide Web site run by the OU Energy and Environment Research Unit: http://EERU-WWW.open.ac.uk.

Rocky Mountain Institute, a centre set up by Amory and Hunter Lovins, provides information on a range of environmentally sustainable technologies, particularly in relation to energy conservation and increased energy efficiency.
RMI is at 1739 Snowmass Creek Road, Old Snowmass, Colorado 81654, USA. Tel: 303 927 3851 Fax: 303 927 4178. RMI can also be contacted electronically at: http://www.rmc.org.

Centre for Alternative Technology, the UK's main public demonstration centre for renewable energy and allied green technologies. It also provides information on do-it-yourself approaches and has an extensive range of publications and a journal *Clean Slate*. CAT, Machynlleth, Powys SY20 9AZ. Tel: 01654 702400. Fax: 01654 702782. e-mail: cat@gn.apc.org. WWW: http://www.foe.co.uk/CAT.

Natural Step, is a worldwide campaign aimed at winning commitment to some basic principles of sustainable development. The UK office is Leigh Court, Abbots Leigh, Bristol BS8 3RA. Tel: 01275 373393. e-mail: <natstep@ecos.demon.co.uk>

Worldwatch Institute is at 1776 Massachusetts Ave, NW, Washington DC 20036, USA. It publishes a range of reports and an annual 'State of the World' review. This is also available on computer disc.

WISE, World Information Service on Energy, provides an information service primarily on nuclear power via its regular communiqués and reports. WISE International, PO Box 18185, 1001 ZB Amsterdam, The Netherlands. Tel: +31 20 612 6368. Fax: + 31 20 689 2179. e-mail: <wiseamster@antenna.nl>. WWW: http: //antenna.nl/~wise.

International Institute for Sustainable Development, based in Canada, produces a regular *Earth Negotiations Bulletin* which reports in detail on the progress of the various international negotiations over emission standards, climate protection agreements and general international environmental legislation. Versions of the Bulletin are available free via e-mail listserver (ask to 'subscribe enb' and send to <majordomo@mbnet.mb.ca>) and via the World Wide Web (at http://www.iisd.ca/linkages/). This Web site is also an excellent starting-point for exploring many of the other organisations which are active wordwide on sustainable development.

Appendix III

General reading

In addition to the specific texts mentioned at the end of each chapter, there are several more general books on energy issues which may be useful: notably Gerald Foley's *The Energy Question* (4th edn 1992, Penguin Books, London); Walt Patterson's *The Energy Alternative* (1990, Boxtree, London) and Cleland McVeigh's *Energy Around the World* (1984, Pergamon Press, Oxford).

Barry Commoner's seminal book *The Poverty of Power* (1976, Jonathan Cape, London) is also still well worth reading for a radical perspective on the political economy of energy.

Finally, there are two recent books which would complement this text. Dave Toke's *The Low Cost Planet* (1995, Pluto, London) covers some of the same technological ground as this book, but puts more emphasis on the economic aspects. *The Future of Energy Use* by Robert Hill, Phil O'Keefe and Colin Snape (1995, Earthscan, London) provides more detailed technical coverage of the energy options, including renewables.

References

Allen, G. R. (1991) *The Scramble for Windfarms*, Northern Devon Group of the Council for the Protection of Rural England.

Allen, P. and Todd, B. (1995) *Off the Grid*, Centre for Alternative Technology, Machynlleth.

Anderson, D. (1994) 'Recent and current trends in US renewable energy', NATTA Report, NATTA, Milton Keynes.

—— (1995) 'Renewable Germany', NATTA Report, NATTA, Milton Keynes.

Arnold, L. (1992) *Windscale 1957: Anatomy of a Nuclear Accident*, Macmillan, London.

Association of District Councils (1992) Report dated February 1991, cited in 'Developing wind energy: the planning issues', Report to the National Steering Committee of the Nuclear Free Local Authorities, May 1992.

Baker, C. (1991) *Tidal Power*, Peter Peregrinus/IEEE, London.

Barac, C., Spencer, E. and Elliott, D. (1983) 'Public awareness of renewable energy: a pilot study', *International Journal of Ambient Energy*, 4 (4): 199–211, July.

Barbrook, R. (1995) 'What's wrong with wired', *Casablanca*, Autumn.

Beckerman, W. (1995) *Small is Stupid*, Duckworth, London.

Beder, S. (1994) 'The role of technology in sustainable development', *Technology and Society*, 12 (4): 14–19, Winter.

Blair, I. (1992) 'Green products', in M. Charter (ed.) *Greener Marketing: A Responsible Approach to Business*, Greenleaf Books, London.

Blowers, A. and Lowry, D. (1991) *The International Politics of Nuclear Waste*, Macmillan, London.

Bouda, Y. (1994) 'Japan's new sunshine programme', CADDET Renewable Energy Newsletter, IEA/OECD, 3, ETSU, Harwell.

Braun, E. (1995) *Futile Progress*, Earthscan, London.

British Wind Energy Association (1995) *Best Practice Guidelines*, BWEA, London.

Bruntland, G. H. (1987) *Our Common Future*, Commission on Environment and Development, Oxford University Press, Oxford.

Capra, F. (1982) *The Turning Point*, Wildwood House, London.

Carson, R. (1965) *Silent Spring*, Penguin Books, London. First published in the USA in 1962.

Central Statistical Office (1992) *Social Trends*, 22, HMSO, London.

Clarke, A. D. (1988) 'Windfarm location and environmental impact', NATTA Report, NATTA, Milton Keynes.

—— (1994) 'Comparing the impacts of renewables', *International Journal of Ambient Energy*, 15 (22): 59–72, April.

—— (1995) 'Environmental impacts of renewable energy: a literature review', Technology Policy Group report, Open University, Milton Keynes.

Cohn, M. and Lidsky, L. (1993) 'What now?'. Paper presented to conference on the 'Next Generation of Nuclear Technology', Massachusetts Institute of Technology, Cambridge MA, October.

Commission of the European Communities (1993) *The European Renewable Energy Study* (TERES), CEC DG XVIII, Brussels.

Cooley, M.(1987) *Architect or Bee?*, Chatto & Windus, London.

Council for the Protection of Rural England (1991) Evidence to the Dept of Energy's Renewable Energy Advisory Group, HMSO, London.

Countryside Council for Wales (1992) 'Wind turbine power stations', CCW Report, Bangor.

County Planning Officers Society (1992) Quoted in Report to the National Steering Committee of the Nuclear Free Local Authorities, May 1992.

Davis, J. (1991) *Greening Business*, Basil Blackwell, Oxford.

Department of Energy (1989) 'The Severn Barrage project: general report', Energy Paper 57, HMSO, London.

Department of the Environment (1992) Secretary of State's response to Inspectors' recommendations, quoted in *Daily Telegraph*, 26 March.

—— (1996) *Indicators of Sustainable Development for the United Kingdom*, HMSO, London.

Department of Trade and Industry (1992) Government's response to the Energy Select Committee's Report on Renewable Energy, DTI, 16 July.

—— (1993) *Digest of United Kingdom Energy Statistics 1992*, HMSO, London.

—— (1994a) 'Outcome of the review of statistical methodologies for the compilation of overall energy data', *Energy Trends*, July.

—— (1994b) 'New and renewable energy: prospects for the UK', Energy Paper 62, HMSO, London.

—— (1995) *Digest of United Kingdom Energy Statistics*, HMSO, London.

—— (1996) 'Energy trends', *Statistical Bulletin*, HMSO, London, January.

Deudney, D. (1982) 'Space: the high frontier', Worldwatch Paper 50.

Deutsche Bundestag (1991) *Protecting the Earth*, vol 1 and 2, Deutsche Bundestag, Berlin.

Deval, B. (1988) *Simple in Means, Rich in Ends: Practicing Deep Ecology*, Peregrine Smith Books, Salt Lake City.

Devall, B. and Sessions, G. (1985) *Deep Ecology*, Peregrine Smith Books, Salt Lake City.

Dickson, D. (1974) *Alternative Technology and the Politics of Technical Change*, Fontana, London.

Donaldson, D. and Betteridge, G. (1990) 'Carbon dioxide emissions from nuclear power – a critical analysis of FOE 9', *ATOM*, 400, February: 18–22.

Donaldson, D., Tolland, H. and Grimston, M. (1990) *Nuclear Power and the Greenhouse Effect*, UK Atomic Energy Authority, January.

ECOTECH (1995) 'Potential contribution of renewable energy schemes to employment opportunities'. Study for the Energy Technology Support Unit, ETSU Report K/PL/00109/REP, Harwell.

Eggar, T. (1993) DTI press release, 15 November.

—— (1994) DTI press release, 11 November.

Elgin, D. S and Mitchel, A. (1977) 'Voluntary simplicity: life style of the future?', *Futures*, 11 (4): 200–206.

Elliott, D. (1978) *The Politics of Nuclear Power*, Pluto Press, London.

—— (1989) 'Privatisation, nuclear power and the trade union and labour movement', Technology Policy Group Occasional Paper 19, Open University, Milton Keynes, November.

—— (1992) 'Renewables and the privatisation of the electricity supply industry: a case study', *Energy Policy*, 20, (3): 257–68, March.

—— (1995) 'The UK wave energy programme', Technology Policy Group Occasional Paper 27, Open University, Milton Keynes, March.

—— (1996) 'Renewable energy policy in the UK: the limits of the market approach', Technology Policy Group Occasional Paper 29, Open University, Milton Keynes, March.

Energy Technology Support Unit (1982) 'Strategic review of renewable energy technologies', vol. 1, ETSU Report R13, Harwell.

—— (1993a) 'Tidal stream energy review', ETSU Report T/05/00155/REP, Harwell.

—— (1993b) 'Attitudes towards windpower: a survey of opinion in Cornwall and Devon', ETSU Report W/13/00 354/038/REP, Harwell.

—— (1994) 'An assessment of renewable energy for the UK', ETSU Report R 82, Harwell.

England, G. (1978) 'Renewable sources of energy–the prospects for electricity', *ATOM*, 264, UKAEA, October: 270–72.

Evans, D. L. (1991) 'Dyfi windfarm worries', *Rural Wales*, Campaign for the Protection of Rural Wales, Spring.

Eyre, B. (1991a) 'The longer term direction for the nuclear industry', *ATOM*, 411, UKAEA, March: 8–14.

—— (1991b) 'The way forward: nuclear power post 2000 in the UK', *ATOM*, 416, UKAEA, September: 9–12.

Eyre, N. (1993) 'Comparison of environmental impacts of electricity from different sources', *Nuclear Energy*, 32 (5): 321–6, October.

Farinelli, U. (1994) 'A setting for the diffusion of renewable energies', *Renewable Energy*, 5, part 1: 77–82.

Flood, M. (1991) *Energy Without End*, Friends of the Earth, London.

Freeman, C.(1992) *The Economics of Hope*, Pinter, London.

Fridleifsson, I. (1996) 'Present status and potential role of geothermal energy in the world', paper presented to the World Renewable Energy Congress IV, Denver, 15–21 June, Proceedings vol. II, pp. 34–9, Pergamon, Oxford.

Friends of the Earth (1994) *Planning for Wind*, FoE, London

Galli, R. (1992) 'Structural and institutional adjustments and the new technological cycle, *Futures*, October: 775–88.

Garrison, J. (1980) *From Hiroshima to Harrisburg*, SCM Press, London.

Genus, A. (1993) 'The political construction and control of technology: wave power renewable energy technologies', *Technology Analysis and Strategic Management*, 5 (2): 137–49.

Gipe, P. (1995) *Windpower Comes of Age*, Wiley, Chichester.

Greenpeace (1992) Evidence to the House Of Commons Select Committee on Energy, hearings on Renewable Energy, Session 1991–2, Fourth Report, vol. II, pp. 157–67.

—— (1993) 'Towards a fossil free energy future', Stockholm Institute report for Greenpeace International, London, April.

—— (1995) Evidence on the CEC Green Paper to the House of Lords Select Committee on the European Communities, Session 1994–5, 17th Report 'European Union Energy Policy' vol. II, Evidence, p. 103, HMSO, London, July.

Greyson, J. (1995) 'Taking a natural step', *New Ground*, 45, SERA, London.

Grob, G. (1994) ' Transition to the sustainable energy age', *European Directory of Renewable Energy*, James and James, London.

Grubb, M. (1991) 'The integration of renewable electricity sources', *Energy Policy*, 19 (7): 670–88, September.

Harper, M. (1994) 'Wind energy still blowing strong', *Safe Energy*, 99: 13, February/March.

Harper, P. (1974) 'What's left of AT', *Undercurrents*, 6: 35–8.

—— (1994) 'The "L" Word: AT and lifestyles', paper presented to the EERU 'AT 2000' Conference, Conference Proceedings, Energy and Environment Research Unit, Open University, Milton Keynes.

—— (1996) 'Getting through to Sweden', *Clean Slate*, 20: 10–11, Centre for Alternative Technology, Alternative Technology Association, Machynlleth.

Herring, H., Hardcastle, R. and Phillipson, R. (1988) 'Energy use and energy efficiency in UK commercial and public buildings up to the year 2000', Energy Efficiency Series 6, HMSO, London.

Hohmeyer, O. (1992) 'Renewable energy and the full cost of energy', *Energy Policy*, 20 (4): 365–75, April.

Horwood, T. (1992) Quoted in *Windpower Monthly*, November.

Hoyle, F. (1980) *Energy or Extinction: The Case for Nuclear Energy*, Heinemann, London.

Hughes, P. (1993) *Personal Travel and the Greenhouse Effect*, Earthscan, London.

Hunt, D. (1991) Planning Consent letter and Annex (P62/546), Welsh Office, Cardiff, 19 September.

Hyam, P. (1995) 'Talking energy', *New Review*, 26, Department of Trade and Industry, August.

Intergovernmental Panel on Climate Change (1995) Second Assessment of Scientific–Technical Information Relevant to Interpreting Article 2 of the UN Framework Convention on Climate Change.

Jackson, T. (1992) 'Renewable energy: summary paper for the renewable series', *Energy Policy*, 20 (9): 861–83.

Jeffrey, K. (1995) *Independent Energy Guide*, Chelsea Green, Vermont.

Jenkins, T. and McLaren, D. (1994) *Working Future? Jobs and the Environment*, Friends of the Earth, London.

Johansson, T., Kelly, H., Reddy, A. and Williams, R. (1993) *Renewable Energy: Sources for Fuels and Electricity*, Earthscan, London.

Jones, P. M. S. (1993) 'Can nuclear power compete?', *Nuclear Energy*, 32 (5): 327–33, October.

Kahn, H. and Simon, J. (1985) *The Resourceful Earth*, Blackwell, Oxford.

Karnoe, P (1990) 'Technological innovation and industrial organisation in the Danish wind industry', *Entrepreneurship & Regional Development*, 2: 105–23.

Keen, B. E. and Maple, J. H. C. (1994) *JET and Nuclear Fusion*, JET Publications Group, Culham.

Kelly, K. (1995) *Out of Control: The New Biology of Machines*, Fourth Estate, London.

Key, C. (1994) *Daily Telegraph*, 27 February.

Kinsella, C. and Bond, M. (1985) 'Solar houses in London', NATTA Report, NATTA, Milton Keynes.

Korton, D. (1996) *When Corporations Rule the World*, Earthscan, London.

Krause, F., Bach. W. *et al.* (1989) 'Energy policy in the greenhouse', International Project for Sustainable Energy Paths, El Cerrito, California, for Netherlands Ministry of the Environment, vol. 1.

Legget, J. (1991) 'Energy and the new politics of the environment', *Energy Policy*, 19 (3): 161–71, March.

Lovelock, J. (1979) *Gaia: A New Look at Life on Earth*, Oxford University Press, Oxford.

—— (1988) *The Ages of Gaia*, Oxford University Press, Oxford.

McKenzie, D. (1991) *Green Design: Design for the Environment*, Lawrence King, London.

Macpherson, G. (1995) *Home Grown Energy*, Farming Press, Ipswich.

McSorley, J. (1990), *Living in the Shadow*, Pan Books, London.

Maisseu, A. and Delanoe, A. (1995) 'Energy in Europe and in the world', *International Journal of Global Energy Issues*, 8 (1–3): 6–30.

Mann, D. E. (ed.) (1981) *Environmental Policy Formation*, Lexington Books, Lexington.

Marx, K. (1974) *Capital*, Lawrence and Wishart edition, vol. 1. Original German edition 1867.

Meadows, D. H., Meadows, D. L., Randels, J. and Behrens III, W. W. (1972) *The Limits to Growth*, Earth Island, London.

Medvedev, Z. (1990) *The Legacy of Chernobyl*, Basil Blackwell, Oxford.

Mellor, D. (1982) Parliamentary Answer, 31 March.

Mensch, G. (1979) *Stalemate in Technology*, Ballinger, Cambridge, MA. Original edition 1975.

Meyer, N., Benestad, O., Emborg, L. and Selvig, E. (1993) 'Sustainable energy scenarios for the Scandinavian countries', *Renewable Energy*, 3 (213): 127–36.

Mitchell, C. (1995) ' REVALUE' Project Outline, Energy Programme Newsletter, 29, SPRU, University of Sussex, Brighton.

Moore, J. (1980) At the opening of the Southampton wave test tank, Department of Energy press release, 26 September.

Mortimer, N. (1990) 'The controversial impact of nuclear power on global warming', NATTA Discussion Paper 9, NATTA, Milton Keynes, September.

—— (1991) 'Energy analysis of renewable energy sources', *Energy Policy*, 19 (4): 374–85, May.

NALGO (1990) 'Greenprint for action', National and Local Government Officers Union/National Extension College course pack.

National Audit Office (1994) *The Renewable Energy Research, Development and Demonstration Programme*, NAO, HMSO, London.

NATTA (1993a) *Renew*, 85, September–October.

—— (1993b) 'The Windfarm Debate', evidence to the Welsh Affairs Committee, vol. III of the Committee's report (WE52).

—— (1994a) 'Story of ugliness of wind turbines: wind turbines the truth,' unsigned Factsheet dated 1993, reprinted in *Renew*, 87, NATTA, Milton Keynes, January.

—— (1994b) 'Windfarm backlash', compilation of press cuttings, vols I and II NATTA, Milton Keynes.

—— (1994c) DTI press release on national wind power programme, quoted in *Renew* 87, NATTA, Milton Keynes, January.

Nieuwenhurst, P., Cope, P. and Armstrong, J. (1992) *The Green Car Guide*, Green Print, London.

Norgard, J. S., Nielsen, P. S. and Viegand, J. (1994) 'Low electricity Europe', in A.T. de Almeida *et al.* (eds) *Integrated Electricity Resource Planning*, Kluwer Academic Publishers, Lancaster, pp. 399–417.

O'Connor, J. (1991) Capitalism, Nature, Socialism, (CNS) Conference Papers, CES/CNS Pamphlet 1, Centre for Ecological Socialism, Santa Cruz.

Office of Science and Technology (OST) (1995) 'Technology foresight: progress through partnership', vol.13, Energy, OST, HMSO, London.

Olivier, D. (1992) 'Energy efficiency and renewables: recent experience in mainland Europe'. Energy Advisory Associates Report, Hereford.

—— (1996) 'Energy efficiency: key to a sustainable energy future', NATTA Technical Paper 15, Milton Keynes.

Owen, G. (1995) 'Energy policy, the government and the energy regulators: a case stduy of the Energy Saving Trust', CSERGE Working Paper GGEC 95–35, Centre for Social and Economic Research on the Global Environment, UCL/UEA.

Peake, S. (1994) *Transport in Transition: Lessons from the History of Energy*, Royal Institute of International Affairs, London.

Pearce, D. (1992) 'Corporate responsibility and the environment', report published by British Gas.

Pearce, D., Markandya, A. and Barbier, E. (1989), *Blueprint for a Green Economy*, Earthscan, London.

Perrow, C. (1984) *Normal Accidents: Living with High-Risk Technology*, Basic Books, New York.

Porteous, A. (1992) 'Energy and waste', NATTA Technical Paper 7, NATTA, Milton Keynes.

Read, P. R. (1993) *Ablaze: The Story of Chernobyl*, Secker and Warburg, London.

RENEW (1996) 'The earth centre', *RENEW*, 100, NATTA, Milton Keynes, March–April.

Renewable Energy Advisory Group (1992) Report to the DTI on Renewable Energy, Energy Paper 60, HMSO, London.

Renner, M. (1991) 'Jobs in the sustainable economy', Worldwatch Paper 104, Worldwatch Institute, Washington.

Ridley, M. (1994) *Daily Telegraph*, 6 February.

Ross, D.(1995) *Power from the Waves*, Oxford University Press, Oxford.

Ryan, C. (1994) 'The practicalities of ecodesign', in M. Harrison (ed.) *Ecodesign in the Telecommunications Industry*, RSA Environmental workshop 3–4 March, London.

Safe Alliance (1992) 'Biofuels and European agriculture', Sustainable Agriculture, Food and Environment (SAFE), London.

Salter, S. (1981a) Quoted in the *Financial Times*, 18 May.

—— (1981b) 'Wave energy: problems and solutions', *Journal of the Royal Society of Arts Proceedings*, London, August.

Scheer, H. (1994) *A Solar Manifesto*, James & James, London.

Schmidheiny, S. (1992) *Changing Course*, Business Council for Sustainable Development, MIT Press, Cambridge, MA.

Schoon, N. (1989) *Independent*, 10 July.

Schumpeter, J. A. (1939) *Business Cycles*, McGraw-Hill, New York.

Select Committee (1984) House of Commons Select Committe on Energy, Session 1983–4, Ninth Report, 'Energy R, D&D in the UK', HMSO, London.

—— (1992) House of Commons Energy Committee, Session 1991–2, Fourth Report, 'Renewable Energy', vol. III, March, HMSO, London.

Shell (1995) 'The evolution of the world's energy system 1860–2060', Shell International, London.

Stirling, A. (1994) 'Diversity and ignorance in electricity supply investment', *Energy Policy*, 22 (3): 195–216, March.

Stoddard, W. (1986) 'The California experience', paper to the Danish Wind Energy Association Conference, quoted in P. Karnoe, (1990) 'Technological innovation and industrial organisation in the Danish wind industry', *Entrepreneurship and Regional Development*, 2.

Thayer, R. and Hanson, H. (1988) 'Wind on the land', *Landscale Architecture*, March.

Thorpe, T. (1992) 'A review of wave energy', ETSU Report R-72, Energy Technology Support Unit, Harwell.

Tickell, C. (1994) 'Concepts and dilemmas of sustainable development', in N. Steen (ed.) *Sustainable Development and the Energy Industries*, Earthscan, London.

Tindle, S. (1996) 'Jobs and the environment', Socialists Environment and Resources Association, London.

Toke, D. (1995) *The Low Cost Planet*, Pluto Press, London.

Trainer, T. (1995) *The Conserver Society*, Zed Press, London.

TUC (1989) 'Green charter', Trades Union Congress, London.

Tylecote, A. (1991) *The Long Wave in the World Economy*, Routledge, London.

United Nations (1992) Agenda 21 Press Summary, United Nations Conference on Environment and Development, Department of Public Information, UN, New York.

Wainwright, H. and Elliott, D. (1982) *The Lucas Plan: A New Trade Unionism in the Making*, Allison and Busby, London.

Walker, S. (1993) 'Down on the windfarm', NATTA Report, NATTA, Milton Keynes.

Wallace, D. (1996) *Sustainable Industrialisation*, Earthscan, London.

Watson, W. (1994) 'Inegrated tidal power', NATTA Technical Paper 10, NATTA, Milton Keynes.

Welsh Affairs Committee (1994) Session 1993–4, Second Report, 'Wind energy', vol.1, HMSO, London, July.

World Energy Council (1994) *New and Renewable Energy Resources: A Guide to the Future*, Kogan Page, London.

—— (1995) 'Energy for our common world – what will the future ask of us?', Conclusions and Recommendations, 16th WEC Congress, Tokyo.

Young, J. (1993) 'Global network: computers in a sustainable society', Worldwatch Paper 115, Worldwatch, Washington.

Index

Note that page numbers for tables are in *italics* and for figures in **bold**.